# DIY RC Airplanes
# from Scratch

## About the Author

**Breck Baldwin** has been building and flying model aircraft since he was a child. He founded the Brooklyn Aerodrome in 2005 to support flying art, education, and technology developments around remote-controlled aircraft. Breck authored a cover story for *Make* magazine, Volume 30, featuring the Flack. He designs custom art planes for festivals, corporate events, and fun. He has led Flack building sessions at summer camps, schools, and various institutions. Breck has a Ph.D. in computer science from the University of Pennsylvania and is the president and founder of LingPipe.

# DIY RC Airplanes from Scratch

## The Brooklyn Aerodrome Bible for Hacking the Skies

Breck Baldwin

Mc Graw Hill Education

New York   Chicago   San Francisco
Lisbon   London   Madrid   Mexico City
Milan   New Delhi   San Juan
Seoul   Singapore   Sydney   Toronto

McGraw-Hill Education books are available at special quantity discounts to use as premiums and sales promotions or for use in corporate training programs. To contact a representative, please visit the Contact Us page at www.mhprofessional.com.

**DIY RC Airplanes from Scratch:
The Brooklyn Aerodrome Bible for Hacking the Skies**

1 2 3 4 5 6 7 8 9 0   DOC/DOC   1 9 8 7 6 5 4 3

ISBN   978-0-07-181004-3
MHID      0-07-181004-8

This book is printed on acid-free paper.

| | |
|---|---|
| **Sponsoring Editor**<br>Roger Stewart | **Copy Editor**<br>James Madru |
| **Editing Supervisor**<br>Stephen M. Smith | **Proofreader**<br>Paul Tyler |
| **Production Supervisor**<br>Pamela A. Pelton | **Indexer**<br>Jack Lewis |
| **Acquisitions Coordinator**<br>Amy Stonebraker | **Art Director, Cover**<br>Jeff Weeks |
| **Project Manager**<br>Patricia Wallenburg, TypeWriting | **Composition**<br>TypeWriting |

This book is dedicated to Robert Cooper, who inspired me with his beautiful airplanes as a child, and my 7th-grade science teacher, Bob Parsons, who taught me how to fly and finish what I started.

# Contents

# Introduction

Welcome to the world of do-it-yourself (DIY) remote-control airplanes. In this book you will learn how to build your own motorized model aircraft from scratch using materials and tools that are widely available and relatively inexpensive. You'll also learn the skills you need to get your plane into the air and keep it there. By the end of the book, you should be able to create your own customized designs, limited only by your imagination . . . and by the laws of aerodynamics.

## Book Overview

This book starts off with very detailed instructions on how to get our Flack ("flying" + "hack") delta wing provisioned, understood, built, flown, and repaired in six chapters that break down as follows:

- Chapter 1 presents a shopping list and possible sources of gear, including a Brooklyn Aerodrome kit.
- Chapter 2 describes the parts of the Flack and what the parts do.
- Chapter 3 details building the deck. The deck houses all the components that control and power the Flack in a tough, resilient form. When the foam airframe is too floppy to fly, then the deck is simply removed and attached to a new airframe.
- Chapter 4 covers building an airframe from foam sheeting.
- Chapter 5 gets you flying. I am surprised by and delighted with how well the flying chapter works—I was not sure that one could learn to fly from text and photos. Some aspects of flying are fundamentally hard, such as turning, but this chapter helps you to ease into the skill one crash at a time.
- Chapter 6 offers checklists, repair diagnostics, and crash kits that will keep you flying.

There will be supporting materials on the book website (http://brooklyn aerodrome.com/bible) that fill out with video what the text and photos struggle with. If you get stuck, fire off an e-mail, but try to figure it out on your own for a while first—you will learn more. Please build a Flack and send a picture to bible@brooklynaerodrome.com. It makes my day every time I get an e-mail with a successful build.

Chapters 7 through 12 should come with the warning sign, "Welcome to the Deep End." This is where discussions are more high level and you are expected to do your own research, fill in the blanks, and generally act like the DIY hero that you are. If this was a cookbook, then the recipes would be more like suggested dishes with some guidance on pitfalls and approaches.

Chapter 7 comes direct from the art department: How to make planes look good during the day. Lots of things have been tried, and some work better than others. This is my assembled knowledge for day fliers.

Chapter 8 covers the natural domain of the Flack—the night sky. In it some of the more effective lighting techniques are presented with a how-to level of detail.

Chapter 9 is a further extension of the art department into novel shapes that fly. It is an excellent place to get familiar with the kinds of shapes that I have gotten to fly. Some fly very well, some not very well, and some are still in development, but the chapter establishes some directions you might want to consider.

Chapter 10 covers very basic aerodynamics, which should ease the path toward creating novel designs.

Chapter 11 is titled, "Hack the Flack: Make and Fly Your Own Design." Welcome to the very deep end. This chapter presents two approaches to prototyping a flying wing–based shape and maximizing the chances of success.

Chapter 12 is a cleanup chapter that has information about autostabilization systems, hacking your own buddy box (student driving for pilots), and flight simulators and general overviews about video, first-person-view (FPV) flying, and sophisticated autopilots.

So that's it. I did the best I could to put everything down that I know. There is still more, but I did try to write what I would have appreciated when I started this crazy project in 2005. I'll tell you about why this whole thing exists below.

## A Brief History of the Brooklyn Aerodrome

Story time—because people ask. As with many things, the Brooklyn Aerodrome and this book came to be via a trail of happenstance, luck, and bloody-minded dedication to something a little crazy. It starts with my "day job" company moving to the Williamsburg section of Brooklyn, New York, to escape the rabbit warren–like office we operated out of in midtown Manhattan. A fellow flyer, Mark Burnham, invited me out to his neighborhood to fly hand-launch gliders in McCarren Park, which sits on the border of Williamsburg and Greenpoint. McCarren Park is a rough, scrabble-surfaced area with three baseball diamonds, that looked like "a pure slice of heaven" to the flying-starved president of Alias-i/LingPipe—a.k.a. me. On the walk back to the subway, I saw a rental sign for offices. Two months later, I moved into a 900-square-foot loft one block from the park with views of the Williamsburg Bridge, the Empire State Building, and more.

At this point, a single table and supplies occupied 50 square feet of the loft in the darkest spot of the office next to the bathroom. I then made the acquaintance of the artist Philip Riley, and we hatched a plan. The plan was to create a night-flying model airplane for the Burning Man Arts Festival in 2005. It was to be electric-powered and feature a masonic eye meant to mock my company's participation in certain Defense Department programs. Until then, I knew next to nothing about electric airplanes, and my struggle in acquiring such knowledge has significantly informed the content conveyed in this book. The name "Brooklyn Aerodrome" came to be because my portion of the project needed a name to write on the shipping containers—really.

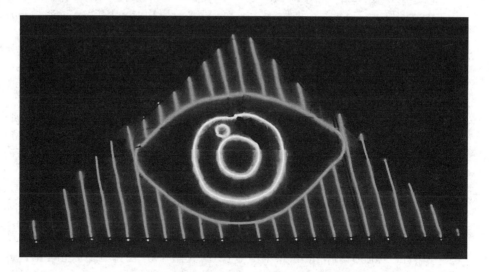

The Flying Eye was a drama-driven plane that absolute rocked people's minds and wrecked Philip and me in the process. It was vastly underpowered, so anger, frustration, and pure will kept it in the air for two nights. I send special thanks to Jewels for the "You gotta do it" Girl Scout speech that humiliated me into finishing the project and flying. Years later, people still recall the Flying Eye, which made it clear that flying art was worth messing with more.

In 2008, after some failed night-flying projects at Burning Man, I went big. The pitch was as follows:

> We are creating an above-ground aeroquarium of illuminated night-flying electric airplanes for Burning Man 2008—August 25 to Sept. 1.
> We need artists to do the designs, pilots, and builders.

The plan was to get 50 night fliers into the air at the same time. The project lead to the genesis of the design philosophy behind the Brooklyn Aerodrome—cheap, quick to build, tough, and flies great! This objective was so stated because I figured that it was easier to get 50 "Burners" to build/fly airplanes than to get 50 accomplished model airplane fliers to go to Burning Man. Burning Man attendees are not likely to have much money or an attention span longer than four hours, and they are going to crash a lot as they learn to fly in the windy conditions of McCarren Park and Burning Man.

In support of this greater effort, the aerodrome grew to 100 square feet, and it was then that the deck was invented—thanks, Splinter. It was then that Rounder came up with the first name for the 42-inch craft we were building. One night Rounder remarked that the delta wing I was about to launch was so beat up it looked like a wet towel. The Towel name stuck in my heart, with the rest of the world hating it, from day one. I thought the textile theme was hilarious and invented the Bib (trainer configuration with autostabilization), the Napkin (small in size), and the Crystal Towel (polycarbonate clear night flier covered later in the book). With the book looming, a contest was announced for renaming the plane, and Ed Gedvila and his twins, Maxwell and Madeline, came up with Flack. Flack is a portmanteau of "flying" and "hack." Much better.

The Aeroquarium was a complete failure at Burning Man. We had six planes but never had more than two in the air at the same time. But we had lots of fun, we learned a lot, and people loved our simple night fliers. Most significant, though, was that the Flack had been born.

In 2009, effort went to doing other arty things with the Flack at the Figment Arts Festival on Governor's Island in New York City and moving to a separate studio space with Mark "Splinter" Harder in Williamsburg and a trip to Burning Man with our now arts person, Karen King.

The studio commitment led to two projects in 2010 that cemented where Brooklyn Aerodrome finds itself today. The first big project was teaching 15 kids at the Beam Summer Camp how to build and fly Flacks over the course of a week. This set us down the road to education. The other significant event was participation in the World Maker Faire at the New York Hall of Science that fall. Our flying was a big hit, and makers loved the Flack. We got lots of media coverage, an editor's choice award, and an understanding that people really liked what the Brooklyn Aerodrome did.

In 2011, the momentum was sustained with an Esplanade theme camp at Burning Man, another teaching experience at Hayground Camp in the Hamptons, and a second World Maker Faire. At the end of the year, I wrote the cover article for *Make* magazine (Volume 30).

In 2012, the *Make* magazine article was published, we participated in the Bay Area, Detroit, and World Maker Faires, and Flack kits went into production for Maker Shed. On the education front, we moved away from directly teaching kids how to build Flacks to teaching teachers how to do it with The Dalton School and 14 sixth-grade girls in June. It was a huge success.

As 2013 closes, the Brooklyn Aerodrome finds itself finishing this book, developing a full education curriculum around aeronautics, and producing kits on a regular basis.

*Breck Baldwin*

# Acknowledgments

In no way, shape, or form did this book get created by just me. Piles of people pitched in, lent gear, and volunteered time, and they must be acknowledged. I live in fear that I have overlooked someone and I apologize in advance for the oversight.

I did something that apparently most DIY book writers do not—test the book with builders. In a very real sense, I am not the best person to write this book because I don't have the perspective of a novice. Testing with novices solved a big chunk of that problem. My testers revealed huge gaps in exposition, ferreted out ambiguity, and made clear which figures were necessary. This book works because of them—it was not working before they showed up.

My novice testers were Joseph Legros, Andrew Woodbridge, and Ed Gedvila and his twins Madeline and Maxwell, for Chapters 1 through 5; they built with no figures other than the plans. Later testers were Rob Tocker, Allen Lubow and his daughter Dana, and David Pollack and his son James (a five-hour build!). They got to have figures, which actually introduced problems. I cut the number of figures in half, and my editor breathed easy that this was not going to be a 600-page book.

A few brave souls ventured into later chapters, saving the reader countless hours of confusion and frustration. Matthew Malina took on the night flying chapter and greatly improved the el-wire illumination sections over the course of a few afternoons. Nate Siegel teaches aeronautics at Bucknell and took on the scribblings of the basic aerodynamics in Chapter 10—all remaining flaws are mine since I didn't take all of his advice. Chapter 11, "Hack the Flack," was tested by Andrew Woodbridge on the heels of his completion of Chapter 5 as a complete beginner with a flying Llama. Ben Bottner and Lowell Weisbord took "Hack the Flack" and designed and built their own flying wings that flew great after a few iterations. They were 12 and 13 years old! Thanks to their parents, Jane Bottner and Joe Weisbord, for all the support.

The book owes thanks to all the fine folks who modeled for the images. Josh Silverman is both the angsty builder and happy flyer in Chapter 2, Bethany Sumner is our flyer in Chapter 5, and Andrew Woodbridge allowed his actual first flights to be recorded for the book.

It is a misnomer to describe the process of creating this book as "writing." In pure hours and psychological terror, the photography and figures of this book dominated the book gestation process. The photography requirements were overwhelming. Shooting for black and white is a totally different game than color, and I would have floundered deeply without the help and equipment of the serious professionals, Aaron Fedor and Philip Riley. Thanks to my father, also a pro, who sent an old Leica prototype digital camera that shot really well. Jimmy Serkoch shot most of the photography for Chapters 3 and 4 and did a great job. Finally, Krishna Dayanidhi, in addition to being a builder, loaned me his very nice gear and lenses for the months it took to shoot the book.

There are some major personalities who made Brooklyn Aerodrome possible. Mark Burnham is the entire reason that the Brooklyn Aerodrome is in Brooklyn. Way back, historically speaking, Robert Cooper and Bob Parsons inspired me to pursue model airplanes as a child. Splinter, a.k.a. Mark Harder, has made huge contributions to the evolution of the Flack and how we operate. Early adopters and supporters include Rounder, Kit a.k.a. Chris Niederer, Bill Suroweic, SuzQ, and Britelite. Jay Kempf does our kits and has evolved what we do to the next level with his long history in model aviation. Rob De Martin from Maker Shed, Steven Gerborich a.k.a. Gerbo, and Hulot a.k.a. Ron Todd have helped and continue to help the Brooklyn Aerodrome make its way. On the business side there are great lawyers, Havona Madama and Rob Griffitts—yes, we need lawyers—and my constantly amused and patient accountant, Jim Dale. Thanks to Dustyn Roberts of *Making Things Move* fame who answered random e-mails from a nervous author.

Organizations that have supported Brooklyn Aerodrome include Kostume Kult, Figment, Container Camp, Nyrvana, Image Node, and especially Maker Faire and *Make* magazine, which got us into print in Volume 30 which led to this book. The Bergen County Silent Fliers offered early and constant support for the Flack—thanks to Carlos Molina, Chris, Ed, and the rest of the club. Most of the first draft was written at Mugs Ale House in Williamsburg—thanks to Scott, Cristina, John, Heidi, Caitlyn, Joanna, Nina, and Drew for asking how it was going and keeping the Flower Power flowing.

Thanks to the McGraw-Hill team of my ever-patient editor, Roger Stewart, and book manager, Patty Wallenburg, who also is a professional handholder for newbie book writers.

This book would not exist without my muse Karen, my wife and the driving force behind art in this organization. It is pretty simple: no Karen, no book. She goes where no one has reliably gone before—next to me in a hot/cold/rainy tent in some forsaken field talking to people about airplanes.

# DIY RC Airplanes
# from Scratch

# CHAPTER 1

# Get Your Stuff

My goal is getting you building and flying your own remote-control (RC) plane as quickly, economically, and easily as possible. This goal entails buying tools, buying parts, and most important, learning to fly. Ordering over the Internet normally takes a few days or weeks if you order from overseas. Assembling your aircraft takes some time as well. By far, learning to fly requires the most time. But you can cheat the time to learn by starting with computer software flight simulators while you're waiting for tools and parts to arrive. There are some very good flight simulators available, either as free downloads or from commercial publishers. I recommend Flying Model Simulator (FMS) or Charles River Radio Controllers Simulation (CRRCsim) of the free-download variety—there is more information on this in Chapter 12. There are also simulators for Android and iOS devices. For about $20, you should get a USB faux RC controller for its standard joystick interface and ease of use. Have at it, and remember to send me a picture or video of what you come up with at bible@brooklynaerodrome.com.

## Introducing the Flack

The basic flyer that you will learn to build from scratch is one we at the Brooklyn Aerodrome call the *Flack*, which is short for "flying" + "hack." The standard form of the Flack can be seen in Figure 1-1. The Flack is designed to be easy to build, cheap, and durable because—trust me—you will crash it many times.

1

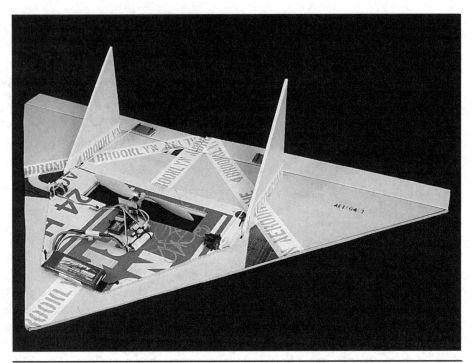

**FIGURE 1-1**    The Flack ("flying" + "hack").

The Flack moniker was created by the family of Ed, Maxwell, and Madeline Gedvila in a renaming contest. It is a huge improvement over the old name the Towel.

Once you have mastered the Flack, you'll be able to extrapolate the design to all sorts of other fun shapes, including the Flying Heart (Figure 1-2), the Manta Ray (Figure 1-3), and the Crystal Towel night flyer made of polycarbonate (Figure 1-4). All the planes use the same basic concepts and skills. Your imagination and the laws of aerodynamics are the only limits to what you can build and fly.

My goal in the first five chapters is to get you flying quickly and inexpensively for around $150, on average. Chapter 1 handles shopping, Chapter 2 addresses what the parts do, and Chapter 3 is a very detailed step-by-step guide to building the deck, or soul, of the Flack. Chapter 4 handles creating an airframe that you will destroy again and again using the same deck to power. Chapter 5 teaches you how to fly. As of this writing, I have had five test builders who ranged from techies to parents with twins work through the instructions. They suffered so that you don't have to. On completion of Chapter 5, you will have the skills to create completely novel aircraft that fly in day or night, which is what the rest of the book is about. The end result is a simple and robust approach to aircraft building. Build it, fly it, and send me a picture.

Time to go shopping.

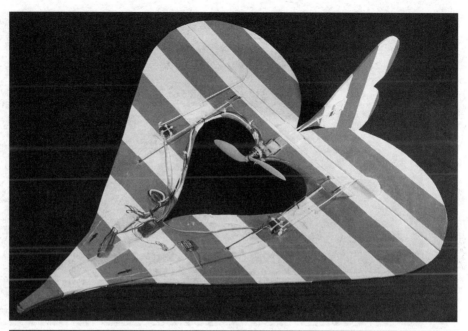

**FIGURE 1-2**    The Flying Heart.

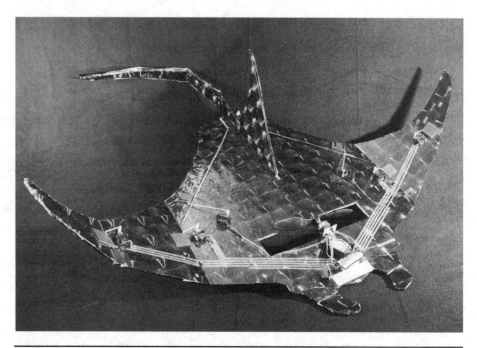

**FIGURE 1-3**    The Manta Ray.

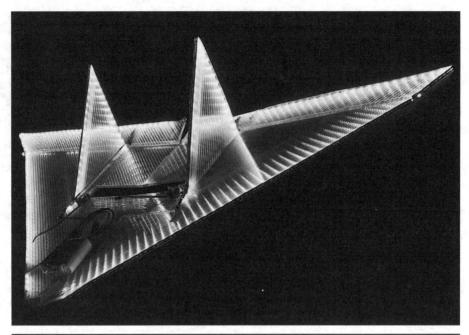

**FIGURE 1-4**   The Crystal Towel.

## Parts and Materials

This section describes all the supplies and parts you will need to build the Flack—tools will be covered in Chapters 3 and 4. I seek to keep costs in check while building a reliable, versatile flyer. In addition to the recommended components, I also point to alternatives that you may want to consider depending on your budget or level of ingenuity. Do not fail to visit the book website for updates on parts, plans, and videos. Kits also will be available on the website.

A checklist of recommended components with sources is given in Table 1-1. A list of recommended spares for frequently breaking parts is included as well. In addition, Figure 1-5 shows the components that are RC-specific. Each part is addressed in greater detail in the following sections with alternatives when appropriate.

### Flight Simulator Controller

Unless you already know how to fly RC, please, please do yourself and the reputation of this book the favor of spending $20 on a USB controller that

**TABLE 1-1** Recommended Component, RC-Specific Parts, and Sources

| Item | Ordered | Have | Description | SKU | Vendor | Cost | How Many | Spares |
|------|---------|------|-------------|-----|--------|------|----------|--------|
| USB Flight Simulator Control | | | Dynam 6 CH USB RC Flight Simulator | 60P-DYU-1002-Simulator | Varied, web search | $20 | 1 | |
| Radio | | | Tactic 4 channel | TTX404 | Varied, web search | $75 | 1 | |
| Motor + prop saver | | | 370 Size Brushless Outrunning 2208 Economy version + Prop Saver 1800kv | N2208 | www.batteryheatedclothing.com | $10 | 1 | 1 |
| Propeller | | | Slow Flyer 9 x 4.7 | GW-EP9047/BK/D | www.batteryheatedclothing.com | $1.25 | 1 | 5 |
| Electronic speed control (ESC) | | | Max 18A Brushless Motor ESC for planes (2A BEC, 2–3 cell Li-Poly) | A 18 A | www.batteryheatedclothing.com | $10 | 1 | 1 |
| 3.5mm bullet connectors | | | n/a | | | | | |
| Battery | | | ZIPPY Flightmax 1800mAh 2S1P 20C (USA Warehouse) | Z18002S20C | http://hobbyking.com | $8.15 | 1 | 2 |
| Battery connector | | | Nylon XT60 connectors male/female | XT60 | http://ebay.com | $3 | 1 | 1 |
| Battery charger | | | Turnigy 2S 3S Balance Charger. Direct 110/240v input (USA Warehouse) | TGY-3 | http://hobbyking.com | $11 | 1 | |
| Servo | | | HXT900 9g/1.6kg/.12 sec micro servo (USA Warehouse) | HXT900 | http://hobbyking.com | $3 | 2 | 1 |
| Foam | | | 24" x 48" sheet Dow HPU plastic/plastic (P/P) | | Hardware store | $3 | 1 | 1 |

*(continued on next page)*

**TABLE 1-1** Recommended Component, RC-Specific Parts, and Sources (*continued*)

| Item | Ordered | Have | Description | SKU | Vendor | Cost | How Many | Spares |
|------|---------|------|-------------|-----|--------|------|----------|--------|
| Motor mount | | | 1 1/2" x 1 1/2" x 1/16" x 1 1/2" aluminum | | Hardware store or old ladder | | 1 | |
| Heat shrink as appropriate | | | 4mm or 3/16" | | Hardware store | $1 | 1" | |
| Hook and loop tape | | | 3/4" wide sticky backed hook and loop tape | | Hardware store | ¢1 | 5" | 5" |
| Small zip ties (.10") | | | .10" x 5 1/2" or longer best | | Hardware store | $1 | 40 | |
| Coroplast at least 11" x 17" 4mm | | | Old signs | | Old signs | | 1 | |
| Double-stick tape, foam tape, etc. | | | Lots of options | | Hardware store | $1 | 15" | 15" |
| 2 wire coat hangers | | | Without paper tube easier to work with | | Cull the herd in your closet | | 2 | |

**FIGURE 1-5**   RC-specific parts.

emulates an RC transmitter. This device, shown in Figure 1-6, allows you to use an RC transmitter-like controller to drive a variety of free simulators for Windows/Mac/Linux. I realize that it blows the budget and is another thing to order, but it will increase the chances of success from 50 to 90 percent in my very subjective estimation. And even better, Flack models are available for a few of the free airplane simulators, as shown in the figure.

Once the controller arrives, go to Chapter 12 to see how to set it up for PC/Mac/Linux. There are lots of flight simulators that cost money, but no need—there are free ones that handle the basics quite nicely once you have a controller.

There are also Android/iPhone/iPad flight simulators available that have touch screen–based joystick emulators that may be of use. Some titles include Leo's RC Simulator and Absolute RC Plane Sim.

## Radios

The radio allows the pilot to communicate control inputs to the receiver, which, in turn, tells the motor how fast to turn and the position the servos should have,

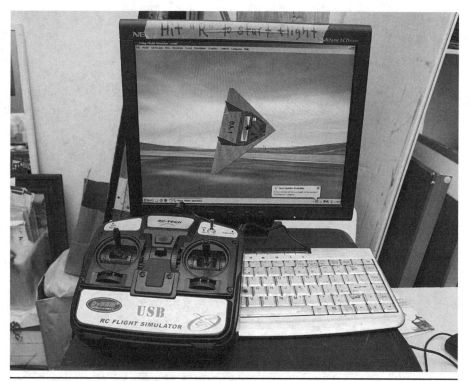

**FIGURE 1-6**    Flight Model Simulator (FMS) with Flack and Dynam USB controller.

and the servos, in turn, adjust the elevons that are responsible for pitch and roll control. Radios have just undergone a major revolution in the adoption of the 2.4-gigahertz (GHz) standard pretty much worldwide. This means that a good deal of older, perfectly good older frequency gear exists that you might be able to get for cheap, but note that the suggested radio is $75 and that there is a $32 option. Receivers employing 2.4 gigahertz are not typically compatible across brands and sometimes even models of transmitters. Be sure to buy a receiver along with the transmitter to avoid this potential problem. All the recommended transmitters come with a matching receiver.

### Modes 1 and 2

In the United States, we fly what is called *Mode 2*, which has aileron and elevator controls on the right stick and rudder and throttle controls on the left stick. It turns out that most of the world flies with a different control setup than the United States, which is called *Mode 1*. Mode 1 has throttle and aileron controls on the right stick and elevator and rudder controls on the left stick. It is rumored that Mode 2 is easier to learn on. Order your transmitter appropriately; otherwise, you will experience extreme awkwardness flying other folks' planes or having them fly yours. This book assumes Mode 2 because I am in Brooklyn, New York, USA.

### Recommended Radio and Receiver

The Tactic TTX404 is a solid radio/receiver at a slightly high price point, but I love it. It is a little odd to program—it feels a bit like a game of twister with the sticks to get elevon mixing working, but all in all, it is a solid piece of gear for around $75. The best feature is a toss-up between the great eco-friendly packaging that protects the radio well and the wireless buddy-box capability. It needs four AA batteries and has a nice neck strap to boot.

### Alternate Radios

The Fly Sky FS-CT6B is my absolute favorite go-to radio for large groups, such as classes and camps. It can be had for $32 new as of this writing from hobbypartz .com. It has allowed hundreds of people to fly Brooklyn Aerodrome aircraft who otherwise might not. It provides a rock-solid radio link at park-flier distances (100 yards/meters) but burns through AA batteries quite quickly and is a hassle to program. There is a vendor who sells alternative software to program the radio if you do a search for "Digital Radio Fly Sky." There are many brands of exactly the same radio with slightly different part numbers, and these include HobbyKing HK-T6A V2, Turborix, Exceed, Storm, CopterX, and Jamara. However, many of these brands do not include the programming cable. Come to brooklynaerodrome.com for programming instructions and the latest availability.

The Hitec Optic 5, 5-Channel 2.4-GHz Sport Radio System is a reasonable radio from a reasonable manufacturer that can be had for around $100 if you look hard. BPHobbies.com is a good source.

The FlySky/iMax/Turnigy 9x/Eurgle 9CH is a $75 radio that has a cult following, with lots of folks developing alternate EPROM upgrades and programming interfaces. I have no personal experience with this, but it looks to be good. If you have every played with an Arduino (a programmable microcontroller), this might be the radio for you.

### Motors

Electric motors now rival internal-combustion motors for power and are much simpler to work with. They don't need gas, don't leave a sticky residue, and are utterly reliable. I use ones that generate lots of thrust at low airspeeds, which means swinging a 9- or 10-inch prop for their size.

There is no standardization around motors that make it easy to shop. In comparison, servos, batteries, and speed controllers are a paragon of clarity. My recommended motor is designated the "370 Size Brushless Outrunner 2208 Economy Version+Prop Saver 1800kV." What does all that mean?

1. "370 Size" is an old designation for brushed motors based on how long the motor housing is in millimeters. Some standard ones are 280, 370, and 400, and those sizes came to be known as a rough designation of motor power. Flacks fly well on 370 to 400 size motors.

2. "Brushless" means that there are no physical contacts required for timing of electricity to the coils of the motor. Old-school electric motors relied on brushes to control which coils got current as the motor turned based on physical contact with the rotating part of the motor. Brushless motors use a sensing technology coupled with a microcontroller to drive the pulsing of energy to the three wires connected to the motor (only two wires are active at any time).

3. "Outrunner" means that the magnets are on the outside with the windings on the inside. This is the reverse of an inrunner. Outrunners are used for higher torque and lower rotations-per-minute (rpm) operation. Inrunners are used in higher-rpm and lower-torque applications.

4. "2208" indicates that the stator diameter is 22 millimeters and that it is 8 millimeters long. See Figure 1-7.

5. "Economy Version" means tough as nails as far as the Brooklyn Aerodrome can tell.

6. "1800kV" means that the motor will turn 1,800 rpm per volt supplied with a small subtractive factor for the resistance of the motor windings. Other factors that determine how fast the motor will rotate include size and pitch of propeller. This is determined largely by how the motor is wound. To provide a rough idea of the calculation, the two-cell batteries I use start at 8 volts, meaning that the motor should spin at 8 × 1,800 = 14,400 rpm.

There is not enough space herein to get into the grisly details of motor specifications, but the preceding should at least lay out the major dimensions behind motor designations. Often the vendors will provide data sheets on thrust, prop size, voltage, and amperes that are pretty conservative on prop size (smaller than possible) and optimistic on thrust.

### Sizing Motors

As a rule, a 1:1 thrust-to-weight ratio for the aircraft helps to deal with typical flying conditions—gusty, windy, limited takeoff and landing areas. It is very useful to be able to power out of trouble in its many forms. Assuming that the aircraft weighs about 1 pound and the battery is a two-cell lithium-polymer

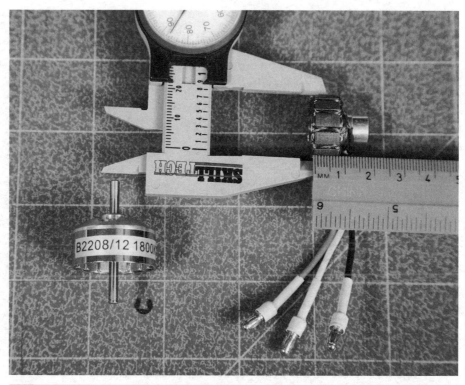

**Figure 1-7**   Stator measurements showing 22-millimeter diameter and 8-millimeter width.

(LiPo) pack, there are lots of options for getting off the ground. Note that suggestions are for single-vendor "combo" deals that match speed controls with motors and include suggested propellers from the same vendor. Stepping outside these combo deals means that you might end up with a speed control with different motor connectors than the motor and/or a prop adapter that doesn't fit the prop. Take care, and check my website for current working combos.

### Recommended Motor–Speed Control–Propeller Combo

RCHotDeals.com (now batteryheatedclothing.com) has been providing motors to the Brooklyn Aerodrome for years. The go-to setup is the economy 370 size 2208/12 motor that runs at 1,800 rpm per volt with an 18-ampere speed control for around $18, as shown in Figure 1-8. It includes a prop saver, which protects the prop by making the attachment slightly flexible. In addition, 3-millimeter bullet connectors are soldered to both the speed control and motor, which is pretty rare and a big time saver. You still have to attach the battery connector, however.

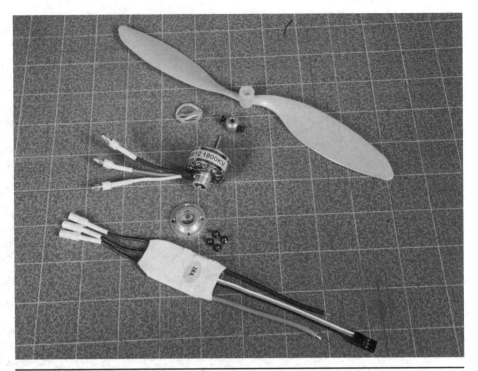

**Figure 1-8**    Recommended motor, electronic speed control, propeller.

The appropriate propeller for beginners is the GWS RD 9x4.7 prop for $1.25, but get at least six because they break all the time. Decoding the propeller designation goes as follows:

1. "GWS" stands for the manufacturer, Grand Wing Servo.
2. "RD" stands for "reduction drive," which in the old days meant a gearbox. The new outrunners have lots of torque at low rpms, so no gearbox is needed but the designation remains. How about "RD" standing for "rpm diminished"?
3. "9x4.7" works out as 9 being the diameter of the prop in inches, and the 4.7 indicates the pitch of the prop. For every turn of the prop, it would advance 4.7 inches if it didn't have any slippage in the air.

I fly with a GWS 10x4.7 propeller with the recommended motor all the time, but it is a size that can easily overtax the motor if run at full throttle for more than a few seconds.

### Alternate Motors–Speed Controls–Propellers

There is a noneconomy version of the 2208 motor that costs a bit more and has an appropriate propeller adaptor for Advanced Precision Composites (APC) style gray props. Figures 1-9 and 1-10 show the combo without a speed control. Note that a circular shim has been used to fit the prop to the prop shaft.

See the section on mounting the propellers in Chapter 3 for more about these issues if you don't get a combo deal that matches propeller, propeller adaptor, and motor. Alternatively, the economy 2212-10 1400kV with a 20-ampere speed control and a GWS 10x4.7 prop is overpowered in exactly the right way, but throttle management will be needed.

Another kind of prop to consider is the GWS DD ("direct drive") 9x5 propeller. It has blades that are slightly thicker at the base and thinner at the tips, but it works fine.

My kits have used BP Hobbies' parts, which are of high quality. The motor is a BP A2208-14 Brushless Outrunner Motor at 1450 kilovolts mated to the BP 18A Speed Control. The propeller adaptor works with APC props, and the 10x4.7 prop works great, but I consider them to be slightly more dangerous because they are sharper, heavier, and stiffer than the GWS versions. The assembly is the same as the non-economy 2208/12 shown earlier. It costs more

**FIGURE 1-9**  The 2208/12 noneconomy motor, prop adaptor, cross-motor mount, and 9x4.7 APC propeller.

**FIGURE 1-10**   The 2208/12 assembled.

but probably produces more thrust. You will need to solder the connectors onto the motor and the battery connectors to the speed control. The nice thing about ordering from BP Hobbies is that the company can get you all the RC-specific parts from a single vendor. The company sells the Hitec Optic 5 radio, the HS-55 servo, and Cheetah battery packs, chargers, and appropriate connectors.

## Batteries

The revolution in motors was accompanied by a revolution in battery technology. The energy density of modern lithium-polymer (LiPo) batteries, combined with the amount of power they can provide at very cheap prices, has opened the door to a new class of aircraft. With this comes a warning: LiPo batteries can catch fire in very dynamic ways from charging, physical damage, or a change in the wind. A great idea is to have a collection of clay flowerpots in which to store and charge batteries. Don't let the batteries get hotter than 120 degrees Fahrenheit or freeze.

A common way to destroy a battery is to leave it plugged in to the speed control when the plane is not being flown. If the battery is discharged too deeply, it will destroy the battery.

At the Brooklyn Aerodrome, we have settled on the 1,800-milliamperehour (mAh) or 1.8-amperehour two-cell battery pack because such battery packs are cheap and are around the correct weight to balance the Flack. For the more creative aircraft, such as the Bat, we double up battery packs for nose weight and will add lead tire weights if needed. It is easier to have a standard size and work around it than to have varying capacities that inevitably will have you at the flying venue with the wrong-sized battery pack. The other considerations around batteries are the speed-control connector and the balancing connector.

---

*CAUTION  These batteries carry a bunch of energy in a small space and can catch fire and explode on overcharging, physical damage, or just plain orneriness. Do not put freshly crashed batteries in your pocket or near anything flammable or burnable. Do not charge them unattended, and store and charge them in a fireproof container— LiPo sacks, a ceramic bowl, a ceramic flowerpot. Figure 1-11 shows new and exploded batteries.*

### Recommended Battery Pack

HobbyKing.com has the least-expensive battery packs historically, and they perform okay. I use very high-quality Thunderpower packs that have not puffed out despite being five years old, but that is the difference between an $8 to $9

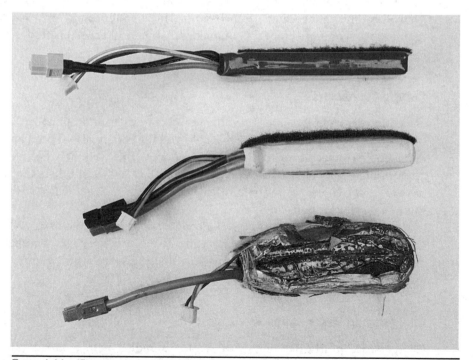

**FIGURE 1-11**    (From top to bottom) New battery, a slightly puffed battery that may still work, and an exploded battery after a crash.

battery and a $30 battery. I suggest going cheap with low expectations. Lots of brands are available, but try to buy from a U.S. warehouse to minimize shipping costs and transit time.

Currently, I use the ZIPPY Flightmax 1,800-mAh 2S1P 20C (U.S. warehouse) at $8 to $9. This battery lasts about one to two years if treated well and has shown itself to be crash-tolerant. It has a JST-XH connector for charging and a male XT60 connector for powering the airplane. Be mindful of these connectors when buying chargers, and you will have to buy and solder a male XT60 connector onto your speed control. You can see this in Figure 1-5 and in Figures 3-24 and 3-25 in Chapter 3.

### Alternative Battery Packs

Any two-cell LiPo pack with a 20C or greater rating will do. BP Hobbies offers the Cheetah Packs 7.4-V, 1,800-mAh, 35C lithium-polymer battery at $16 with JST-XH connectors for charging and Cheetah 4.0-mm controller connectors. Visit the Brooklyn Aerodrome website to see how we got this to work.

Be aware that lithium-iron-polymer (LiFePo) and lithium-iron (LiFe) batteries are now being sold that look very much like LiPo batteries but differ in a few ways. They have a lower voltage (6.6 volts versus 7.4 volts for LiPo), so if you go with these, be sure to go on the higher end of kilovolt ratings for the motor—1,800 kilovolts is about right—and consider using a 10-inch prop where I have been recommending a 9-inch prop. These batteries also will require a charger that is designed for them. A major advantage of LiFePo batteries is that they are less likely to catch fire.

### Battery Connectors

Generally, the battery connector to the speed control is determined by the one soldered onto the battery. The site that sold you the battery also should have the appropriate connector for the speed control on sale as well. Be sure to get the appropriately gendered connector for the XT60, Deans, and others that don't have the same connector for both sides. Powerpoles, EC2, and Cheetah use the same connector for both sides.

I have found that Deans and XT60 connectors can be difficult to get apart. Children in particular can struggle getting connectors apart and actually can damage the airplane when the connector finally gives way. Powerpole, EC2, and Cheetah connectors are easier to take apart.

### Chargers and Power Supplies

Chargers are nearly as complex a world as radios with a broad range of possibilities. As with radios, higher quality comes with higher price and higher complexity. You also have to factor in how the charger is going to be powered—choices are alternating-current (ac) wall power (110/220 volts) or 12- to 17-volt direct current (dc) provided by a car battery or ac power supply. Chargers are the one place where the capacity of the charger is tightly linked to price. If you are dedicated to doing this project for around $150, then the charger will be $11 and will take between 2 and 3 hours to charge a battery. A more expensive charger at $45 will charge in 20 minutes. Even if you have a car that can power the charger, you almost certainly will want a power supply to run the charger in the house.

There are lots of other options out there. There always seems to be a decent charger available from my three choice vendors, Hobbypartz.com, BPHobbies. com, and RCHotdeals.com. Make sure that the charging connector on the battery matches that on the charger. Almost everyone is using the JST-XH connector, but check.

### Recommended Cheap Charger Route

The Turnigy 2S, 3S Balance Charger with direct 110/240-volt INPUT (USA Warehouse) from Hobby King works well for all us urban folks who lack a car because it runs off 110-volt ac. We are pretty disenfranchised here in the city, but really, there is no other way. This charger also runs off 220-volt ac for similarly affected world citizens and is shown in Figure 1-12. The big downside is that batteries will take 2 to 3 hours to charge.

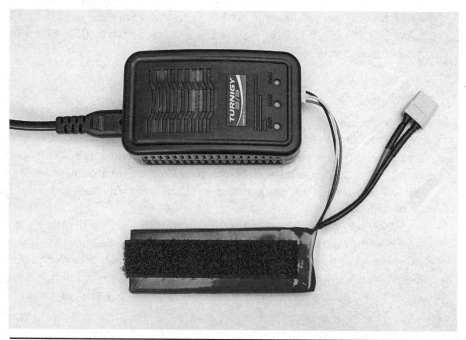

**Figure 1-12**    Charging connector attached to charger.

A bit of quick math on charger capabilities and time to charge is in order. The Turnigy charges at around 600 milliamperehours, or 0.6 ampere per hour, for a two-cell pack. The battery contains 1,800 milliamperehours, or 1.8 amperehours, of energy that, when depleted, needs to be recharged. The Turnigy will recharge at 0.6 amperes per hour, meaning a 3-hour charge for a completely depleted battery. See below for the 20-minute solution, but it is going to cost you.

### Alternative Recommended Charger and Power Supply Combo

My go-to charger is the excellent but fussy-to-program Cellpro Multi4 Charger with GP/KO (JST XH/JST EH/HR) Adapter Combo at fmadirect.com or revolectrix.com ($65). The adaptor should fit the batteries suggested earlier, and the charger will need a 12- to 15-volt power source. Battery manufacturers overestimate risk and recommend charging only at what is called 1C. "1C" means "one times capacity," which for our batteries is charging at 1.8 amperes per hour, which would take 1 hour to charge a completely depleted battery. Some modern batteries are advertised as capable of 2C (30-minute) or 3C (20-minute) charge rates. I play it safe and charge at 2C, and I make sure that I have some spare charged batteries on hand. This charger will let you charge at up to 4 amperes per hour, which is just a bit more than 2C.

### Other Chargers

BP Hobbies offers the simple GWS (Grand Wing Servo) C3-LP Charger (GWCHG004) for $15 and the GWS DC 15-V Adapter for the C3-LP Charger (GWCHG006) at $10. Charge time is 2 to 3 hours. This will allow charging from a car or from a wall socket.

### Power Supplies

Most chargers need a 12- to 15-volt power supply. You can use a regulated bench supply or a wall wart that delivers sufficient amperes for your charger. A good source of high-ampere power supplies is old computers. It takes a bit of effort to extract them, and you have to be comfortable digging around electronics that can kill you dead if you make a mistake. There are many how-to guides on this on the Internet.

Wherever you buy your charger, the vendor generally will offer an ac power supply to run the charger as an additional purchase. Be sure that the ampere requirement of the charger (expressed in amperes) is less than or equal to what the power supply can provide. Most wall warts have about 1 ampere of power at 12 volts—the specification should be listed on the power component.

### Servos

Servos take the signal and power from the receiver and move the servo arm to the position indicated by the transmitter. This, in turn, allows us to control the elevons, which govern pitch (up/down) and roll. Servos are broadly and cheaply available at around $3 to $10 each—get a few extra because they can strip on crashes. If the servo has metal gears, you can expect it to be tougher and less likely to strip gears on crashes.

For the planes we fly, we use the class of servos defined by the canonical Hitec HS-55 ($10). The desired specs are a weight of around 10 grams and torque of 16 ounces per inch, meaning that it can raise 16 ounces 1 inch. You can go with bigger servos if you have them around. There are many options here, so try to find a good deal from either your radio, motor–speed control–propeller, or battery vendor, and add them in—it is hard to go wrong.

### Recommended Servos

The Hobby King HXT900 ($2.69) has nice long servo wires and is a constant favorite. It now features fashionable black gears and servo arms.

### Alternative Servos

RCHotdeals.com (batteryheatedclothing.com) has an 8-gram servo that is two for one at $7.99. BPHobbies.com stocks the Hitec HS-55 at $10. Read and react based on other places where you are buying servos, and take pleasure that these servos used to cost $25 back in the day.

### Airframe Materials

Finally, we are out of the rarified world of RC electronics. All the stuff just mentioned is going to be very hard to hack, recycle, or otherwise create from whole cloth. Some day we will try to create a Flack from totally recycled parts, but that someday has not arrived yet. Now you can get creative. You need just a few major components.

### Motor Mount

The motor mount for some reason is always a difficult point for me. I started with blue plastic gutter-like material that the neighboring venetian blind shop threw out, moved to a worn-out aluminum ladder, and have played with three-dimensional (3D) printed wonderfulness. Just be creative—all that is needed is a 90-degree support for the radial or cross-motor mount that comes with the outrunner and you are good. Examples are shown in Figure 3-8 of Chapter 3.

**Industrial Solution.**  For Brooklyn Aerodrome kits, I use aluminum angle stock that is $1\frac{1}{2} \times 1\frac{1}{2} \times \frac{1}{16}$ inches. You can get this from a hardware store. The problem is that you have to buy 8 feet of it at a time, which is a lot of motor mounts at $1\frac{1}{2}$ inches wide. This is what is used in Chapter 3.

**Aluminum Ladder.**  *Do not cut a motor mount out of a working ladder!* Seriously, you could kill somebody. I found a stepladder that was worn out (wear on the steps, believe it or not), and for about a year all Brooklyn Aerodrome planes featured steps from that ladder as motor mounts.

**3D Printed Motor Mount.**  Technophobe thought that it could improve on the motor mount with a full 3D printed design; see it at www.thingiverse.com/thing:14507. I have one that remains untested because I don't have a project worthy of the effort. I think this reveals why I like to build with trash.

**Coroplast Motor Mount.**  For a while, after having run out of ladder, I built with a Coroplast motor mount that worked really well, except when it got hot and melted. It also required a lot of precision cutting, which is counter to the Brooklyn Aerodrome approach.

## Deck Material

**Coroplast or Plastic Cardboard.**  I hate this stuff, but I love the fact that every time someone builds my beginner plane, a bit of Coroplast stays out of the landfill and brings someone joy—it is the material used throughout Chapter 3 to attach the gear to. Please try to recycle this stuff—my first batch of signs was a bunch of Newport cigarette signs that someone had tossed. A happy end to the political season is the collection of yard signs no matter who wins. I use the 4-mm thickness the most. It is fairly light (2.8 ounces per square foot) and very durable. The two major uses are for decks on the Flack, reinforcement for the art planes, and elevon control horns. It is not a great material for wings and stabilizers because of its weight, potential to cause damage, and its tendency to be a little too flexible, which introduces a control-lockout situation with the elevons.

**Blue Foam.**  The Dow blue fan-fold in Figure 1-13 is my bread-and-butter building material for wings and stabilizers. It is quite light (¾ ounce per square foot) and very stiff for its weight, but it can be very difficult to find. If you do find it, you have to buy a bunch, although I have heard of folks being able to buy just a few sheets. It comes in a fan-fold package of twenty-five 2- × 4-foot sheets, which is sufficient material for approximately 35 to 40 Flacks, and it costs

**Figure 1-13**    Dow HPU ¼-inch plastic/plastic (P/P) ready to fly.

around $50 to $70 a package. If you get the whole bundle, you can think of it as a sketchbook for creating flying things.

The official name is ¼-inch Dow fan-fold high performance underlayment (HPU) insulation with plastic bonded to both sides (P/P). You may need to call Dow to find a source. Do not tell Dow that you want to build airplanes with it because the company will send you to a vendor who did a custom order of white foam that lacks the film on both sides, which adds a great deal of stiffness to the foam.

Note that Dow makes a version with film only on one side, which is less stiff. The foam stiffens up beautifully if faced with wrapping paper or adhesive film—but it also becomes less crash-tolerant.

**Cardboard.**    I have built planes from cardboard in pursuit of a more recycled plane but never really had great success with it. Figure 1-14 shows two efforts in cardboard that flew but not particularly well. The problem is that cardboard is quite heavy, which makes for an airplane that is underpowered with the typical motors. Also, the cardboard is not very damage-tolerant and doesn't easily provide a stiff control surface. But don't let my experiences limit you. Using this material will shift the center of gravity (CG) back. See Chapter 10 for ways to handle this.

**Figure 1-14**   Cardboard, a great idea, needs refinement.

**Pink Foam.**   Corning makes a pink version of the blue foam but in ⅜ inch that I have used in the past. In fact, the first Flack was made of it, as shown in Figure 1-15. It is a bit heavy but adequately stiff and thick enough that you can sand in a bit of an airfoil. Using this material will shift the CG back. See Chapter 10 for ways to handle this.

**Depron.**   Depron is a flooring underlayment available in Europe and elsewhere and by special order in the United States. It also comes in white and lacks the stiffness of the Dow foam, but it is a popular building material. You may need to take measures to stiffen it.

**Foam-Core Board.**   Most art shops carry foam-core board that can be used in the ¼-inch size. It is heavy, stiff, and damage-intolerant. I would be tempted to strip off the paper layers and work with the foam directly. Lots of this stuff gets thrown out every day, so it would be great to figure out a way to "aircycle" it.

**Dollar-Store Foam Core.**   In the New York City area, there are various dollar stores that carry a fairly light but fragile foam-core board. Lowell, who is all of

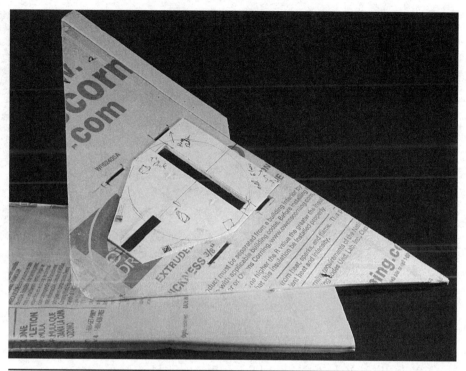

FIGURE 1-15    The first deck on the first Flack.

13 years old, built his first Flack out of it. It doesn't take damage particularly well, but it got Lowell going, which is the most important thing.

## Tape

One of the more unexpected aspects of running the Brooklyn Aerodrome is the amount of experimentation we have done with tape. Every time I wander into a hardware store or an art store, I head over to see what the world is doing with tape. Even simple packing tape is quite a complex world of varying ultraviolet (UV) resistance, adhesive quality, and thickness. Colors can come into play, and Brooklyn Aerodrome even has custom tape. Figure 1-16 shows the kinds of tape we use.

### Packing Tape

Packing tape provides the hinge for the elevons and reinforces the stabilizers and nose of the wing. It also can be used for decoration if colored. A dispenser, even a cheap one, makes the tape much easier to work with. I prefer thinner (2.2-mil) colored tape from a reliable brand such as Uline or Scotch/3M. Others

**Figure 1-16**   (From left to right) Frog tape, foam tape in ¾-inch squares, carpet tape, high-stick tape, Velcro tape, and custom packing tape on the bottom.

can work fine, but I have had bad experiences with elevon hinges failing because of insufficiently sticky tape. I use 2-inch-wide tape.

### Double-Stick Tape

Lots of components need positioning before reinforcing with zip ties. Double-stick tape helps tremendously. It also can add a lot of strength.

Some tapes we have used at Brooklyn Aerodrome are discussed below.

**¾-Inch Foam Tape Squares.**   At the Brooklyn Aerodrome, we use thin foam tape from Uline (¹⁄₃₂ inch thick) in our kits and classes. It speeds up the building process. All sorts of other kinds of foam tape are used for mirror hanging, etc. All should work well.

**Carpet Tape.**   Carpet tape is double-stick tape that is quite sticky and tends to release later without too much fuss. It is also pretty cheap at around $7 a roll. We get 2-inch rolls but have moved away from the material in favor of foam squares.

**High-Stick Tape.**   This tape is strong enough to replace zip ties for servo and elevon control-horn placement. It is just a layer of adhesive as opposed to a paper or foam substrate that has adhesive on both sides. It is very good for sticking to decorated airplanes that have plastic skins and a plastic surface on the deck. It is expensive and hard to remove from components when it sets up. It also can release if it is very cold.

### Velcro Tape

Velcro tape is used for attaching the battery to the airframe. Get the ¾-inch regular-strength variety. Higher grip strength (from the hook/loop side) makes it very difficult to replace the battery because it sticks so much. The adhesive also can release in the cold.

### Frog Tape or Low-Stick Painter's Tape

Frog tape comes in handy when you are positioning parts, packing stuff, or labeling things. My favorite is the green 1-inch Frog tape.

### Cable Ties/Zip Ties

I use varying lengths of 0.10-inch UV-tolerant ties—usually the black ones. You can get 1,000 four-inch ties for about $5 online at Cabletiesandmore.com. Ties can be expensive at hardware stores ($3 for 100). However, the ties can be doubled or tripled to get longer lengths, as discussed in Chapter 3.

## Conclusion

So this is the stuff we use to build at the Brooklyn Aerodrome. Check the website for up-to-date information, and please innovate and share. Chapter 2 goes into what these parts do.

# CHAPTER 2

# What Do All the Parts Do?

This chapter covers the basic concepts of remote-controlled (RC) aircraft as seen through the lens of Brooklyn Aerodrome's quick-to-build, inexpensive, and tough Flack flying wing. This chapter is directed at the complete novice. This is where terms such as *elevon*, *speed control*, and *trim tab* get explained. While the chapter generally should be useful for other kinds of RC aircraft, I am focusing on what is needed to get flat-plate-based flying wings working in your world. There are lots of ways to get flying—this is my way.

## The Flight Cycle of a Flack

Once the Flack is built and the pilot has basic control of the plane, flights proceed as follows:

1. The flight battery is charged. This can take 3 hours on a low-power charger to 20 minutes depending on the capabilities of the charger and battery. It is a good idea to have more than one battery charged.
2. The transmitter is always turned on first with the throttle down.
3. The Flack is powered up by connecting the flight battery via the main power leads. There is no power switch on the Flack. The charging connector is not connected to anything.
4. A quick preflight check verifies that up/down/left/right commands work.
5. The throttle is advanced to two-thirds (depending on conditions).
6. The Flack is hand launched into the wind.
7. The Flack flies for 5 to 8 minutes depending on throttle settings.
8. The Flack should be flyable 1,000 feet away, but cheaper transmitters may not work that far. It is more fun to fly close in anyway. The Federal

Aviation Administration (FAA) requests that model aircraft stay below 400 feet of altitude.

9. The *cone of crashing potential* dictates the pilot's actions at all times—see Chapter 5.

10. Once the pilot knows how to fly, the Flack can do loops and rolls and fly inverted.

11. It can carry 4 to 6 ounces of payload, such as a camera or a parachute drop, or it can tow a streamer.

12. When the pilot notices that 4 minutes have passed or notices that the Flack seems to be lacking power, then he or she should begin to plan the landing.

13. Landing, like takeoff, is optimally done while flying into the wind.

14. The Flack is brought in under reduced power to land, with the last 5 feet or so of altitude being a pure glide with the motor off.

15. The Flack lands on its belly.

16. The pilot should check the motor for overheating by putting a finger on the motor housing.

17. Then repeat these steps with a fully charged flight battery.

## The Pilot

My method starts with a person with good vision (corrected is fine), awareness of his or her surroundings, and a need to fly, as shown in Figure 2-1. If the pilot is a beginner, then he or she should have spent a good deal of time practicing flying on a simulator. Beginners can learn to fly without a simulator, but a simulator really helps with the basic dynamics of flying in a consequence-free environment. Chapter 12 tells where to get free/cheap flight simulators and how to set them up. The pilot is responsible for all that happens regarding the aircraft, even if events are out of his or her control. A healthy dose of worse-case-scenario reasoning is very helpful.

## The Builder

The builder (Figure 2-2) is often the pilot, but different skills are required for each task. The builder is patient and reads all the instructions once all the way through and thinks about what is happening before building. If the builder doesn't understand what a step in the instructions is achieving, then he or she thinks about it until it is clear. Understanding what the goal of a step is makes a huge difference in execution because instructions cannot cover all contingencies. This skill is hugely valuable when planes need repair (they will) and when

**FIGURE 2-1** Pilot and assorted equipment.

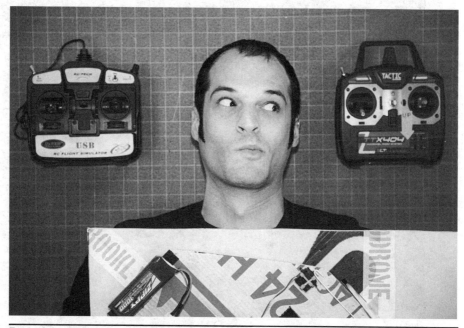

**FIGURE 2-2** Builder and assorted equipment.

innovation is desired. If there are two or more builders, then one should read the instructions out loud and discuss with the other what they are going to do.

## The Airframe

The airframe is where the Brooklyn Aerodrome style begins to express itself. The quick-to-build-from-scratch requirement forces us to work in the realm of flat plates of foam and tape and to minimize the number of parts. A traditional airframe has a separate fuselage, main wing, and tail. This means a lot of cutting, and the end result is not likely to tolerate damage well.

Without further ado, let's get to know the core design of the Flack, as shown in Figure 2-3.

### The Wing

This big triangle of foam is where all the lift comes. The delta shape is a classic flying form that existed long before the invention of supersonic fighters. Alexander Lippisch was designing and flying very similar shapes before World

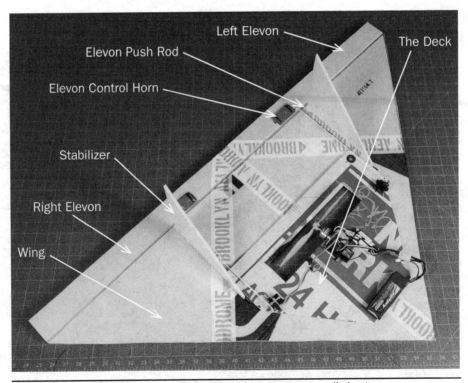

**FIGURE 2-3**    The core design of the Flack with major components called out.

**FIGURE 2-4**  *Alsomitra macrocarpa* seed. (*Photo courtesy of Scott Zona.*)

War II. Nature even predates him with a flying-wing seed that has existed for 56 million years, as shown in Figure 2-4.

### The Elevons

The rear surface of the wing has two control surfaces that handle pitch and roll control. They are independent of each other, and both are used for aileron inputs and elevator inputs. Because both elevator and roll control are integrated into both control surfaces, the surfaces are called *elevons*, which is a portmanteau that works out as "elevator" + "aileron" = "elev + on."

### The Stabilizers

These surfaces function much in the same way as rudders on traditional airplanes or a skeg on a surfboard. But because they have no movable control surfaces, they are called *stabilizers*. They control yaw and keep the nose of the plane pointed straight into the airstream. Chapter 10 provides more detail about the aerodynamics of aircraft.

Flat-plate model airplanes have been around for quite a while. Materials vary from foam board to Coroplast to, my favorite material, Dow fan-fold. We all owe a great debt to those who innovated this approach to building park fliers.

## The Deck

The deck is where all the electronics are attached, which brings me to another key feature of aircraft—revivification.

A brief story: Mark "Splinter" Harder had built an early version of the Flack with all the radio gear attached directly to the Dow blue foam that we currently use at Brooklyn Aerodrome. After a night of learning to fly, his delta wing was destroyed, and he was frustrated that he would have to redo all the work of attaching gear to another airframe. He suggested that the radio gear should be easily removed and reattached to the foam airframe. I brought out a piece of Coroplast that the radio gear could be attached to independent of the airframe, and the modern Flack was born. Figure 2-5 shows the initial drawing on the destroyed airframe from that conversation. The deck (Figure 2-6) does a few things.

1. It allows rapid transfer of the fussy radio gear to a fresh airframe in very little time.
2. It usefully stiffens the foam around the prop hole.

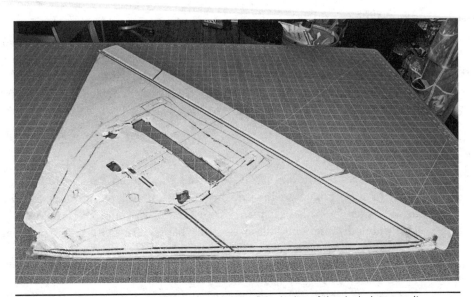

**Figure 2-5**   An early destroyed delta wing with the first design of the deck drawn on it.

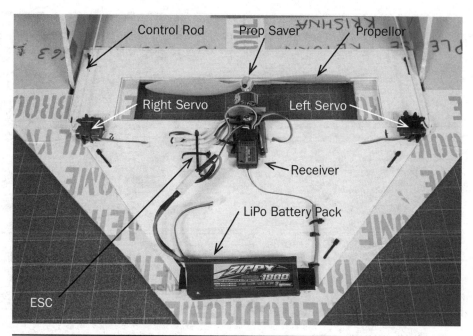

**Figure 2-6** Deck with major components called out.

3. It firmly attaches all the bits that want to fly off in a crash to a very strong substrate via zip ties and/or tape.

Moving on, we then have to control the Flack from a remote position. Enter the transmitter.

## The Transmitter

Figure 2-7 shows a typical transmitter with labeled controls for Mode 2 flight, as used in the United States. The right stick is meant to mimic the traditional yoke of an aircraft with the stick perpendicular to the ground. Yoke control is standardized in full-scale aircraft so that pulling back on the stick raises the nose—think of movie fighter pilot scenes where someone is screaming "Pull up! Pull up." *This is the opposite of typical game controls.* Novice pilots with a healthy dose of experience with games struggle with reversing their notions of up/down control. Try to sort this out on a simulator. This will be covered again in Chapter 5, where the reversal of roll control will be addressed as well.

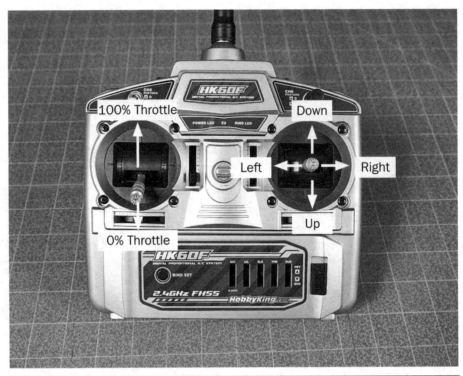

**FIGURE 2-7**    Typical transmitter with Mode 2 controls.

## How the Transmitter Works

The pilot's eyes via the brain connect the pilot's thumb/finger control inputs to the sticks of the transmitter. Those sticks communicate the desired position of both control surfaces and how fast the motor should be turning to the receiver. The underlying technology is not very important to understand, but the result is—if you move a stick on the transmitter 50 percent of its possible travel, then the motor or a servo will respond similarly and hold it. This is an idealized version of what is going on, but it is the general concept—variables include transmitter mixing of controls, programmable end points, etc.

The range of the transmitter should be at least 1,000 feet in clear air, but manufacturer claims vary wildly, and actual performance depends on conditions. Transmitters provide a range-test capability that reduces the power of the transmitter so that you can test the sensitivity of the receiver at smaller distances—consult the transmitter manual. For lesser equipment, the range can be a few hundred feet.

## How to Control the Sticks

There are two major approaches to applying hands to the control sticks—thumb style and pen style. Pen style (control both sticks like a pen) is very popular among precision aerobatic/helicopter flyers, and the thumb style is popular with most everyone else (control sticks with thumbs on top of sticks). It's a personal choice. Note that pen style generally requires a neck strap to support the transmitter's weight.

Pilots should endeavor not to let the sticks autocenter by releasing them and allowing the centering springs to zero the controls. Flight simulators reinforce this behavior by having aircraft that are very stable and do not require constant corrective inputs. The real world is not so forgiving.

## Mixing

On a typical RC airplane layout, there is one servo that controls the elevator, which, in turn, is responsible for pitch control. With a flying wing, both elevons must move up to provide up control. One control must control two servos. How does this happen?

Lots of approaches to this have been cooked up over time, but the modern solution is that the transmitter mixes inputs from the elevator control to the two elevon servos so that both servos respond with 50 percent of their upward/downward direction throw coming from the elevator control. Likewise, the transmitter mixes 50 percent of the aileron input into aileron outputs. Mix ratios may vary based on pilot preference or presets. Consult your radio manual for instructions on how to do this.

The correct elevon to control mapping is shown at the end of Chapter 4.

## Throttle

The throttle is the "gas pedal." Figure 2-7 shows the correct mapping of the throttle. Be careful about running 100 percent throttle other than when you are flying, and even then use full throttle judiciously because Brooklyn Aerodrome designs are overpowered.

## Rudder or Yaw Control

The rudder control (left stick, left to right) is not used in Brooklyn Aerodrome designs and can be ignored. The rudder is usually used to help control the yaw of the airplane.

## Trim Tabs

Notice the labeled trim tabs in Figure 2-8. The trim tab adjusts where the "center" of the controls is located. If you ever had a car that pulled to the left or right when you took your hands off the wheel, in airplane terms, you would adjust the left/right trim tab to change the center of the wheel.

The trim tabs allow for small adjustments to the corresponding control's range. Around 10 percent of the control's movement can be adjusted with the trim tab. If the plane is diving all the time, then the elevator (up/down) trim tab can be adjusted up a bit. Likewise with the aileron (roll). If the plane is always turning left, then you can dial a little bit of right in with the trim tab. In calm conditions, the airplane should be able to fly hands off for 50 yards without a problem. This is achieved by setting the trim correct. The trim tab on the throttle should be all the way down because some speed controls require the maximum negative value for the control to arm and power the motor. The rudder trim tab is irrelevant because Brooklyn Aerodrome designs don't use rudders.

Trim-tab starting points are shown for the aileron/elevator (centered) and the throttle (down) in Figure 2-8. There are also digital trim tabs that use a rocker switch to achieve the same result—consult your radio manual.

**FIGURE 2-8**  Trim tabs.

## Rechargeable Batteries

Higher-priced transmitters come with rechargeable batteries. The less-expensive end of the spectrum uses disposable AA batteries. The cheapest I have ever found for AA batteries is eight for $2, so it doesn't take long for that to add up to more than the cost of a set of rechargeable AA cells.

# The Receiver

The receiver sits on the airplane and takes signals from the transmitter and commands the attached servos/speed control. Figure 2-9 shows a receiver with the major parts called out. The antenna can be very fragile and difficult to replace, so take care that it is well secured on the airplane.

Figure 2-9 shows a lot of wire coming out of a typical receiver in the form of three-wire connectors. Two of the wires are for power, as indicated by negative being black or brown and the positive being red. Positive is always the middle wire. If your receiver came with a manual, then use that to make sure that the three wire connectors have the correct orientation. Sometimes there is an indication of connector orientation on the receiver itself. In the absence of any

**Figure 2-9**  Detailed receiver connections.

information, the negative terminal tends to be toward the edge of a top plug receiver, as shown in Figure 2-9. See the section on the electronic speed control and battery for a discussion of how the receiver gets power to operate.

### How Servos Get Information

Servo motors need power to move, and that power comes from the positive (+) and negative (–) wires. The power from the receiver is around 5 volts, but it can be higher for performance systems. Notice, however, that the Flack has a variable throttle and servos that hold a position—how do they get the information about where to be?

That information is relayed via the signal wire, which is either white or yellow. It conveys a position that by convention ranges over 200 percent. Zero percent is in the middle, –100 percent is one extreme, and 100 percent is the opposite extreme of servo movement. For the throttle channel, –100 percent is no throttle, 0 percent is half throttle, and 100 percent is full throttle.

The expected servo or throttle position is communicated by pulse-width modulation (PWM) that works as follows: Fifty times a second the receiver sends a pulse of voltage that lasts between one-thousandth and two-thousandths of a second. The short pulse indicates to the servo motor controller that it is to have the output at –45 degrees or –100 percent, the long pulse is +45 degrees (100 percent), and 1.5-thousands of a second indicates 0 degrees (0 percent). Figures 2-10 and 2-11 show the maximum extents of servo movement.

The servos do their best to achieve that instruction. Some specialty servos are capable of throws greater than 90 degrees total.

## What the Radio Channels Do

Four- to six-channel radios are what I use in this book, but you will be using only three channels for the standard designs. Channels map to a single signal for a servo or throttle. There are ways to have more than one servo or motor on a channel, but they all will be getting exactly the same signals. The mapping I use in this book is as follows.

### Channel 1

Channel 1 is traditionally the aileron channel, which controls roll. It is used for the right elevon. All left/right/nose/tail/front/back descriptions are from an imagined pilot's point of view facing the direction of flight. A servo is connected to channel 1, which, in turn, is connected to a pushrod that connects to the right elevon control horn—it is not much of a horn, but that is its proper designation. That output arm connects to a rod that connects to a hinged control surface that is called an *elevon*, as shown in Figure 2-12.

**Figure 2-10**    Full right aileron with corresponding servo arm throw.

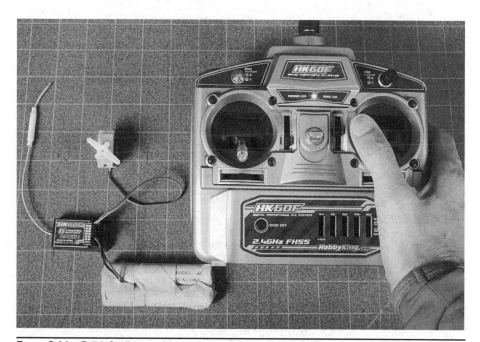

**Figure 2-11**    Full left aileron with corresponding servo arm throw.

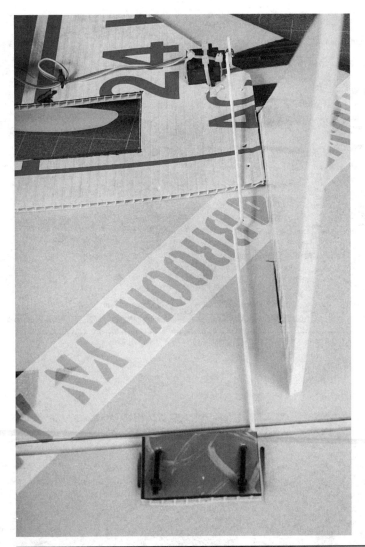

**Figure 2-12** The right servo connects to the control rod, and the control rod connects to the right elevon horn.

## Channel 2

Channel 2 is traditionally the elevator channel, which controls pitch on traditional aircraft, but Brooklyn Aerodrome design uses it for the left elevon.

## Channel 3

Channel 3 connects to the electronic speed controller (ESC), which, in turn, powers the receiver and servos and any other accessories, drawing from the receiver via the positive and negative wires. The signal wire communicates to the speed control how much power to provide to the motor. More about the ESC next.

## Electronic Speed Control and the Battery Eliminator Circuit

Things are getting a little "acronymy," but these are common usages and worth getting familiar with. I use *electronic speed control* (*ESC*) and *speed control* interchangeably. The speed control has two major components: (1) the circuitry to send pulses of energy to the motor so that it turns and (2) the battery eliminator circuit (BEC), which provides power to the receiver and connected items such as servos. Figure 2-13 identifies the inputs and outputs. Note that there are three motor power leads. A very useful property of this setup is that to reverse the rotation of the motor, you switch any two of the three leads.

### Power Limitations of ESC and BEC

The speed control is a good place to address the power limits of the various connecting components. The speed controls we use at Brooklyn Aerodrome have approximately the following specs:

1. The speed control will draw 18 amperes of power from the battery pack at around 7.4 volts (two cells). In reality, the battery is finished charging when it reaches 8.4 volts. That voltage will drop to 8 volts under any

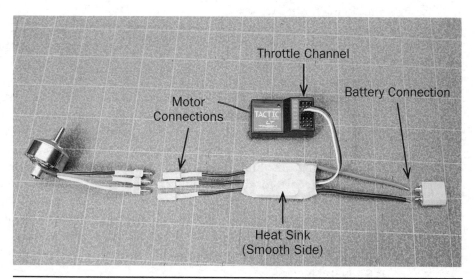

**FIGURE 2-13** Inputs and outputs of the speed control and what they connect to.

serious load, and over the course of a flight it drops to 6 volts. The ESC should limit the discharge to 6 volts while powering the motor. To calculate watts, you multiply amperes times volts. Thus 18 amperes $\times$ 8 volts = 144 watts, which is pretty impressive for such light gear—more than a typical lightbulb's worth of energy. Speed controls will rate the maximum voltage that they can handle, usually expressed in how many LiPo cells for which they are designed. On our end of the scale, you will see two- to three-cell limits.

2. Most of the 18 amperes will go to turning the motor via sequenced ac bursts. Speed controls will be destroyed if a motor draws more amperes than the speed control can handle. The number of amperes a motor will draw is usually described in the motor documentation, and it is also a function of how big a propeller the motor is swinging—see the motor section below.

3. Modern speed controls will cut off the power to the motor at a predetermined voltage while keeping the receiver and servos powered to allow a controlled gliding landing. For two-cell batteries, the cutoff is around 6 volts. Speed controls generally ship with reasonable defaults and will automatically recognize the cutoff voltage, but it is worth looking at the manual to verify this. LiPo battery packs will be destroyed if they are discharged below 3 volts per cell under load. Some speed controls will allow a burst of motor power if you drop the throttle to zero and bring it back up, but use this feature sparingly. The BEC will function as long as possible, even if the motor cutoff has been hit, and can take the battery to less than 3 volts per cell, resulting in destruction of the battery. Do not leave the battery attached to the speed control after flying for this reason.

4. The typical speed control provides 2 amperes of power to the receiver, attached servos, and other power-drawing accessories such as night-flying electronics. Typical sub-microservos draw a maximum of 0.5 ampere and the receiver 0.1 ampere, so 1.1 amperes are needed for the typical two-servo setup. Night-flying gear can easily add 1 ampere, for a total of 2.1 amperes, which may lead to power brownouts on hard maneuvering. Some receivers will reboot on a brownout, which can take 1 to 2 seconds and is more than enough time to result in a crash. In addition, if the BEC limit is exceeded, it may burn out. If the energy budget is exceeded for the BEC, either select a speed control with more BEC power or get a separate BEC from the speed control. Castle Creations, purveyors of very high-quality gear, offers both a 10- and a 20-ampere peak BEC if lots of energy is required. You can disable most speed control BECs by cutting the positive wire if an independent BEC is being used.

Speed controls have a heat-sink side, indicated by a smooth surface, that is meant to be exposed to air for cooling. Install that surface in good airflow.

## The Flight Battery

The flight battery has a simple but important role to play: It powers everything on the airplane. It is the descendant of a laptop battery crossed with a cell phone battery. You can expect it to last between 5 and 10 minutes depending on throttle use. It is also the volatile prima donna on the airframe that can burn your house down if you treat it badly. Refer to the warnings in Chapter 1—they need to be taken seriously. Get at least three batteries and a good charger to keep flying continuously.

## The Motor

The motors we use at Brooklyn Aerodrome are the descendants of CD-ROM motors, and fancier versions are now on Mars driving the Curiosity Rover around. Brushless outrunners have high torque and high efficiency at low weight. This means that no gearboxes are needed, and a very small motor can swing a very big propeller. Repeating again from the speed-control section, to reverse the rotation of the motor, switch any two of the three motor leads. The enemy of motors is heat. It degrades the magnets, and if it melts the insulation on the windings, the motors short out and emit a puff of black smoke, stop turning, and may even destroy your speed control in an instant. Heat builds up from turning a propeller that is too big or from inadequate cooling. Test the motor temperature after every flight by placing a finger on the bell housing for at least 5 seconds. If the motor is too hot to touch, then seek the remedies covered in Chapter 6.

## Conclusion

This chapter maps the equipment listed in Chapter 1 to how it is going to be used on an airplane. Before moving on, get on a first-name basis with all the parts of your airplane, and know what they do. It is time to start building a Flack. Chapter 3 covers building the deck, and Chapter 4 covers creating the airframe.

# Building the Deck

Finally, the time has come to build the plane. But first you need to build what is called the *deck*, which takes the most time and holds/protects all the finicky electronics. The deck is the soul of the Flack because the deck lives many lives via airframe replacement. Make it well, and make it strong. Read this entire chapter before starting. Have all your parts on hand. It should take a few afternoons. What needs to be done accurately will be made clear, and there is not much of that. Relax, enjoy the build, and worry not—you can always start over. As an aside, spend some time with a flight simulator to get the basics of flying under your belt if at all possible.

This is fun and doable, but realize that you are going to be needing some possibly unfamiliar skills. This is not Legos. You are building an airplane from scratch, and there are lots of places to make mistakes. Be patient. You will fly, and we will back you. Contact us at the Brooklyn Aerodrome for help (bible@ brooklynaerodrome.com).

Complete build videos are available at http://brooklynaerodrome.com/bible. You are strongly recommended to view them. Some concepts are much better explained with video. Information about kits is there too.

## Tools and Supplies

Tools and supplies are listed below and shown in Figure 3-2. You probably can make do with less. The following are needed:

1. An Exacto-type knife or a box cutter to cut Coroplast
2. Needle-nose pliers to cut and pull zip ties

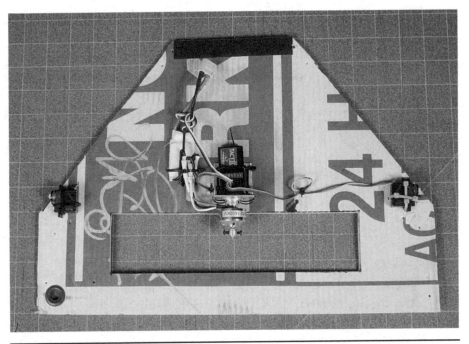

**Figure 3-1**  A finished deck.

3. If you are using a prop saver:
   a. A medium Phillips head screwdriver
   b. A rotary tool with a grinding wheel or a hand file
4. A small Phillips head screwdriver for punching holes in Coroplast and securing servo arms in Chapter 4
5. A drill with assorted bits ($\frac{3}{32}$ inch, $\frac{1}{4}$ inch, 9 millimeter)
6. A drilling surface with a C-clamp
7. A 1.5-millimeter hex key for the motor mount, if needed (Hex keys are extruded hexagonally shaped rods bent into an L shape.)
8. A hacksaw or other means to cut the motor mount from aluminum
9. A soldering iron with solder and "helping hands" ("Helping hands" are clips that hold small parts for soldering.)
10. A heat gun or lighter
11. A ruler
12. Vise-Grips or another way to clamp things
13. A fine-point felt-tip marker
14. Fine/medium-grit sandpaper for deburring aluminum
15. Alcohol or acetone solvent for cleaning servos of grease
16. A workspace that is 48 × 36 inches and tolerates scratches from being cut on

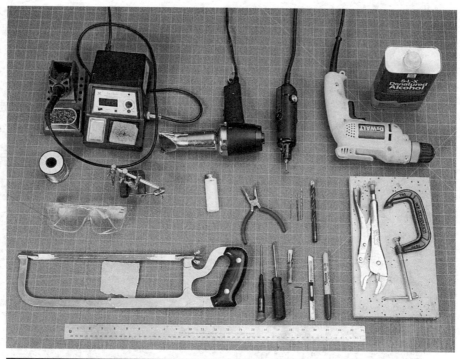

**Figure 3-2**  Tools for the deck build.

17. A releasable thread lock compound such as Loctite (The thread lock acts as glue on the threads of a screw to keep it from unscrewing as a result of vibration and other forces. It is not absolutely required, but be sure to check screws for loosening.)

18. Safety glasses to protect your eyes from blobs of hot solder, flying metal, and other hazards

Parts shown in Figure 3-3 include

1. Transmitter/receiver
2. Brushless outrunner motor with prop saver or prop adaptor
3. One or more propellers (three recommended), either GWS RD 9×4.7, GWS Direct Drive 9×5 or 10×4.8, or APC 10×4.8 (Be sure that the props are compatible with your prop adaptors or prop savers.)
4. An 18-ampere electronic speed controller (ESC) with battery eliminator circuit (BEC)
5. Appropriate 3.5-mm motor/speed-control connectors as needed

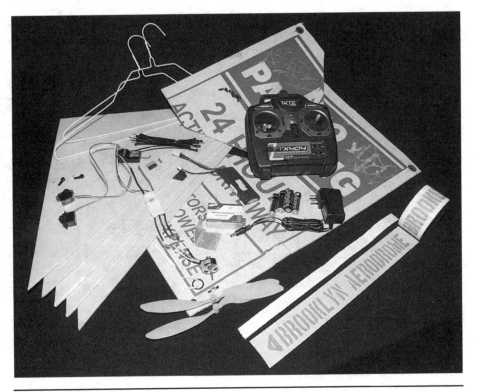

**Figure 3-3**    Flack parts.

6. An 1,800-milliamperehour two-cell 7.4-volt lithium-polymer (LiPo) battery (Be careful of fire possibility, which is most likely with physical damage. Do not store/charge batteries in or on flammable materials.)
7. Battery connector for speed controller
8. LiPo battery charger with power supply if needed. *Read the instructions before using.*
9. Two servos
10. Motor mount (aluminum shown)
11. Hook and loop tape
12. 4-mm or $\frac{3}{8}$-inch heat-sink tubing
13. Small (0.10-inch) zip ties, 4 inches or longer—at least 40 if 4 inches, 30 if 5 inches
14. Deck: an 11- × 17-inch × 4-millimeter recycled Coroplast sign
15. Double-stick tape (Carpet tape is very good or double-sided foam tape.)
16. Two coat hangers (Chapter 4)
17. Foam (Chapter 4)

## Charge Your Battery

Charging your battery is the first step so that it will be ready for powering the deck at the end of the chapter. Depending on chargers, it can take up to 3 hours to charge a battery from empty. LiPo batteries typically ship 70 percent charged. Depending on your charger, you may need to make up connectors for the power leads of the battery—see the section on soldering the speed control for instructions on soldering.

## The Deck

The deck is where most of the effort happens by attaching servos/motors, etc. to the Coroplast with a combination of tape and zip ties. None of the build is particularly precise, but it is good to make sure that servos are well attached, wires are not going to get drawn into the prop, and the motor mount is solid. I have decks that have been on 10 or more airframes and still fly. I like to recycle decks as much as possible. The deck used for this build was found on the street in Brooklyn.

### Fabricate the Deck

Trace the plans shown in Figure 3-4 onto the deck material, and cut it out. The flute direction is not important. The prop cutout should be 1 inch bigger than the prop. If you are a beginner, then I recommend a 9-inch prop. You can always make the hole bigger later, but beginners benefit from having extra strength in the deck/wing and the more limited power of the 9-inch prop.

Take care when cutting, and make sure that you have a sharp knife. Multiple light passes with the knife guided by a ruler works well. Save excess material for elevon control horns and for a backing surface when punching holes in the deck and airframe.

### Attach Servos

The servo wires must reach the appropriate connections on the receiver. Looking from back to front—the front of the deck is the 5-inch narrow part—the right servo goes into channel 1 and the left servo goes into channel 2 unless your radio manual indicates otherwise. The receiver will be placed later, but understand the constraint that the servo wires should cross by at least 1 inch when the servos are attached to the deck.

**Figure 3-4**   Deck plans.

**Figure 3-5**   Punching under mounting flange.

The servos can be shifted left or right to accommodate short leads, but be sure that the rear of the servo (the side opposite from where the servo wire comes out) is aligned with the front edge of the prop hole. This edge aligns the deck and the airframe, as well as the length of the control rods. The servos can be as far inboard as the edges of the prop hole. Have the servo wire toward the front and output shaft toward the outside.

Connect the servos solidly to the deck. If they move on a crash or hard landing, it will change the trim of the airplane and make it harder to fly. I use double-sided carpet tape to place the servos and reinforce it with crossed zip ties. The tape is not mandatory, but take extra care to get the zip-tie holes next to the servo body so that the servo will not wiggle. Take extra care even if you use tape. Steps include

1. Remove servo labels on the side to be attached to deck. If they are very difficult to remove, then don't worry about them—just be sure that the double-stick tape has a solid connection to the servo.
2. If the servos are greasy, clean them with some alcohol or acetone.
3. Apply double-stick tape to the deck side of the servo.
4. Place the servo on the deck. Check Figure 3-1 to be clear about placement and orientation. The important placement is that the rear edge of the servo be aligned with the front edge of the prop cutout. Also double-check that the servo leads overlap by 1 inch at least. The servo output shaft and three-conductor wire should be oriented toward the front.
5. Punch closely to the edge of the servo with a thin screwdriver, as shown in Figure 3-5. The punch locations are (1) next to mounting flange on both sides and (2) next to gear housing on top of servo and (3) the corresponding point on the bottom. Look at the figure. Use some scrap Coroplast under the deck to punch into—it helps to make sure that both sides of the deck end up with holes in them.
6. Reinforce with crossed zip ties as shown in Figures 3-6 and 3-7. The blocky zip connection needs to be on top of the deck to allow flush mounting to the airframe later. If you have 4-inch zip ties, you will likely need to double them to wrap the entire servo. Five-inch zip ties are long enough. You do this by
   a. Running a zip tie through the two holes for the servo in the deck.
   b. Attaching the blocky/zippy part of another zip tie to the end of the one you just put through the deck.
   c. Now there should be enough length to wrap the servo, as shown in Figure 3-7.

**Figure 3-6**    Crossed zip ties at bottom of deck.

**Figure 3-7**    Doubled and crossed zip ties at top of deck.

## Create the Motor Mount

Figure 3-8 shows various motor mounts that have been used for our airplanes at Brooklyn Aerodrome. Plastic motor mounts are a bit more flexible in crashes and tend to be easier to fabricate but can get soft when the motor gets hot. People have melted off Coroplast motor mounts by running the motor indoors in a hot room for a few minutes. In extreme cold I have had plastic motor mounts snap. Metal motor mounts do a better job of dissipating heat and hold up pretty well. They can bend on hard crashes, but I have never broken one. I have made metal motor mounts out of ladders and angle stock from hardware stores. An aluminum can–based motor mount was attempted, but it was not stiff enough.

Aluminum stock is a very easy-to-work-with metal that can be treated like very dense wood when cutting and drilling. Regular drill bits will work fine. It can be helpful to use a punch or nail to indent the aluminum before drilling to keep the bit from wandering. Cutting is best done with a fine-toothed hacksaw or similar blade.

Be creative when looking for material. If need be, you can go to the hardware store and get aluminum angle stock that is $1\frac{1}{2} \times 1\frac{1}{2} \times \frac{1}{16}$ inches, but it comes in

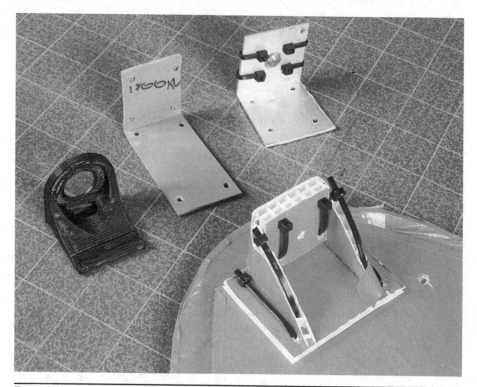

**FIGURE 3-8**   Various efforts at creating motor mounts.

long lengths usually. Use a hacksaw to cut it to a width of 1½ inches. Use medium-grit sandpaper to smooth off any sharp bits—this is called *deburring*.

## Motor-Mount Interface with Deck

These instructions assume that the angle-stock or ladder option is being used. Drill four holes through the motor mount for eventual mounting to the deck, and the holes should be slightly bigger than the zip ties (I use ⅛-inch holes for 0.10-inch-wide zip ties). One detail worth attending to is to stagger the holes by ¼ inch fore/aft next to the 90-degree angle, as shown in the deck plans. This minimizes a stress riser on crashes that can bend the motor at that point. The forward holes are not subject to those stresses.

## Motor-Mount Interface with Motor

Somehow the motor has to be attached to the motor mount. When attaching the motor, be sure to have the power leads off to the side of the angle mount and not over the top or under the deck. Instructions for three common options follow.

**Option 1: Circular Motor Mount.**   This motor mount has the advantage of making the motor slightly more replaceable at the flying site. The motor is held with an Allen set screw, and replacement involves little more than unscrewing the set screw and disconnecting the bullet connectors. I use 0.10-inch-wide zip ties to attach the circular mount to the back of the angle stock or other versions of the same. Screws could be used as well, but they tend to vibrate loose, so I avoid them. If you use screws, then use releasable thread-lock compound to keep them in place.

The circular motor mount can be used as a guide to drill the angle-stock motor mount. Figure 3-9 shows the two mounts being drilled with a ³⁄₃₂-inch drill bit, which is just a bit smaller than the zip ties that keep everything nice and tight. You may need to remove the Allen set screws to access the holes cleanly.

Circular motor mounts tend not to have sufficient clearance for the propeller shaft. The mount should be drilled to allow proper seating of the motor, as shown in Figures 3-10 and 3-11.

When attaching the circular motor mount, be sure to keep easy access to the Allen set screws or removal of the motor will be difficult to remove once it is attached to the deck. Remember to replace the Allen set screws if you took them out for drilling.

Finally, attach the motor to the motor mount with wires off to the left or right side of the mount. Be sure to tighten the Allen set screws and apply thread-lock compound.

**Figure 3-9**  Circular motor mount being used as a drill guide.

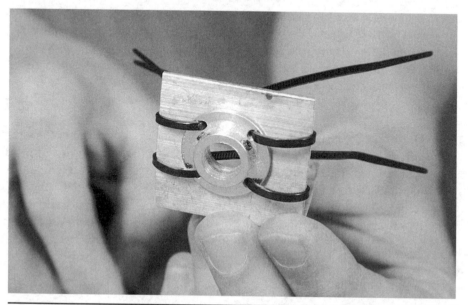

**Figure 3-10**  Circular mount attached to angle stock. Note the easy access to the Allen set-screw heads at 11 and 1 o'clock.

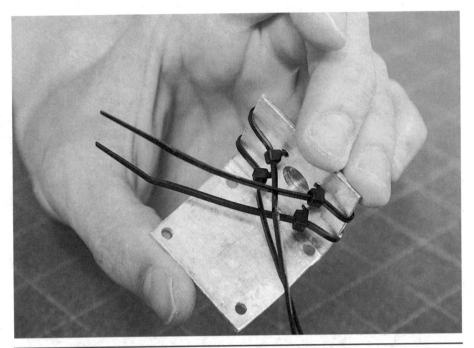

**Figure 3-11**    Back of angle stock where zip tie heads are located.

Tightening screws into aluminum is a delicate process. Too loose and the screw doesn't hold tight; too tight and the threads are stripped. My approach is to tighten until I am sure that I am almost done and give an additional one-sixteenth turn or so to make it tight. You should really feel the force necessary go up. Be careful. Applying a little releasable thread lock is a good idea, too.

**Option 2: Cross-Mount.**    Cross-mounts differ from circular motor mounts in that they use four screws to attach a cross shape to the motor that, in turn, has four holes to attach to the angle stock. Use the cross-mount as a template to drill the angle stock. Then attach the motor to the cross-mount via the included screws. Note that the screws tend to be asymmetrically spaced, with one set 19 mm apart and the other 22 mm on the motor-to-cross-mount connection. Use of releasable thread lock is a good idea. No shaft clearance is needed.

### Mount Propeller

Next up is attaching the propeller to the prop shaft of the motor. There are three main ways to get a prop mounted to a motor, as shown in Figure 3-12.

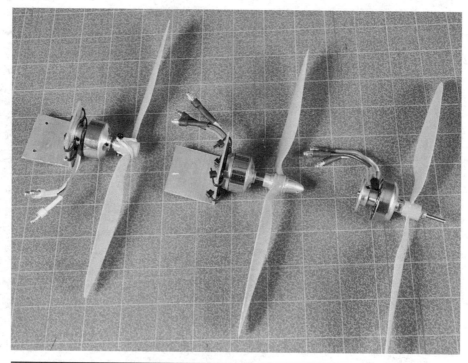

**Figure 3-12**   (*Left to right*) Option 1: Prop saver; option 2: collet-style prop adaptor; option 3: direct screw on prop with nuts.

### Option 1: Prop Saver

Prop savers are designed to make the propeller have some give when it hits ground or another object, which hopefully protects the prop and the motor. This is a good idea that may or may not work. Steps are as follows:

1. File a flat spot on the prop shaft to help prevent the set screw from slipping (Figure 3-13). I use a Dremel rotary bit to do this, but a file would work, too. I grind the flat spot the length of the shaft and take off enough material that I can feel the flat spot with my finger ($\frac{1}{32}$ inch). It is also helpful to grind a mark on the bell housing so that you know where the flat spot is.
2. Screw out the set screws that also hold the prop on so that the prop saver can slip over the motor output shaft. Verify that it can slide onto the shaft—there may be flakes of aluminum getting in the way.
3. Fit the prop to the prop saver. The raised writing on the propeller should be facing the prop saver. Note that the prop saver usually has two different sizes. Pick the one with a snug fit.

FIGURE 3-13　Grinding flat spot on prop shaft for better retention of prop saver—wear safety glasses.

4. Many motor shafts are bigger than the GWS propeller hole. If so, slide the prop saver down until the motor shaft stops it going further, and then tighten the set screws using thread-lock compound. One screw will rest on the flat that you created earlier, as shown in Figure 3-14. A drop of releasable thread lock compound should be used on the screws.

5. Attach the prop with a rubber band or O-ring. Figure 3-15 shows both methods.

6. If you are using APC propellers, the hole is correctly sized or one of the included adaptors hopefully will fit. You may need to drill out the hole on the propeller to get a snug fit.

### Option 2: Propeller Adaptor

The propeller adaptor provides a rigid connection between the motor and propeller and is typical with APC-style propellers because of the large propeller hole used. As in the prop saver case, it is a very good idea to buy the adaptor with the motor to be sure that the motor shaft size fits.

Attachment is more straightforward:

1. Put the collet onto the motor shaft. It should fit snugly. If it is too tight, you need to buy a different adaptor. If it is too loose by just a little, you can add a wrap of tape to make up the missing diameter of the shaft.

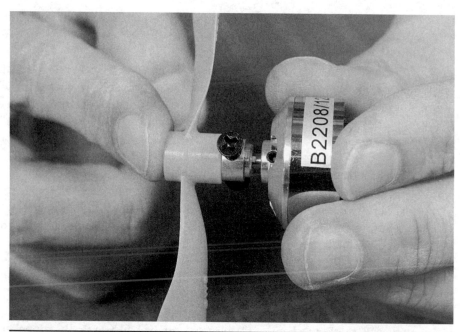

**FIGURE 3-14**   Note that the prop saver does not slide all the way down. There should not be a gap between the prop saver and the prop. A gap between the prop saver and motor bell housing is fine, as shown.

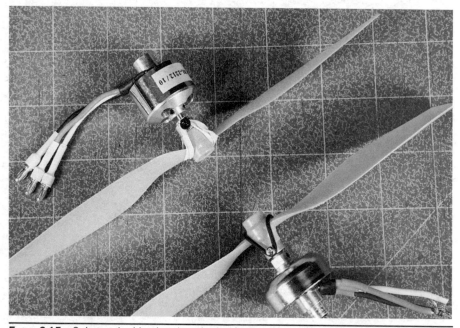

**FIGURE 3-15**   O-ring and rubber-band methods of attaching the prop to the prop saver. Hair holders work well too.

2. Place the writing on the prop facing the motor, and add the tightening screw to the threads.

Prop adaptors have a terrible habit of unscrewing themselves on landing if the power is left on. Always check that the adaptor is still on when you pick up your plane so that you don't lose the adaptor in tall grass. APC props come with adaptor shims that may help to fit the adaptor shaft. Read the enclosed instructions.

### Option 3: Threaded Motor Shaft

Some motors, such as the Tower Pro 2408-21T, have threaded motor shafts that can interface directly with GWS-style props. I have found that on landing they tend to spin the securing nuts off the shaft, so take care on crashes and landings with power on. Figure 3-16 shows the install.

### Attach the Motor Mount to the Deck

There are two ways to approach aligning the motor mount on the deck: (1) Get out the rulers and markers, or (2) just use your eyeball. Center the propeller left-right in the prop hole, and align the front edge of the motor's bell housing on

**Figure 3-16**   Threaded propeller interface.

**Figure 3-17** Place the back of the motor bell housing on the front edge of the prop hole. Center the prop left to right.

the front edge of the prop hole, as shown in Figure 3-17. An angle-stock motor mount will be set back from the edge of the prop hole a bit. Feel free to use your eyeball—it will come out fine. Steps include

1. Attach the motor to the motor mount if not already done.
2. Attach double-stick tape to underside of the motor mount.
3. Determine placement of the motor mount with the goal of proper orientation of the propeller.
4. Stick and reinforce the motor mount with zip ties, as shown in Figure 3-17.

## Solder the Speed Control

Figure 3-18 shows speed controls in need of varying accoutrements. Two ends of the ESC need attention—the battery connection and the motor connection. Start with the motor connection, which pretty universally involves 3-millimeter bullet connectors. My favorite vendor solders these connectors beforehand, leaving only the battery connection to be resolved. Almost all other vendors leave the motor connections up to the builder. Below are instructions for soldering bullet connectors, followed by some options regarding battery connections and how to solder them.

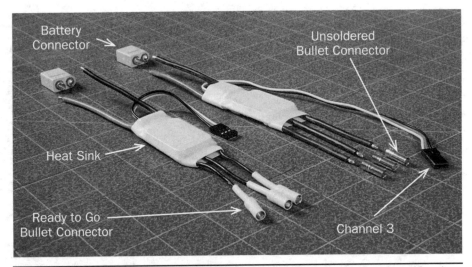

Battery
Connector

Unsoldered
Bullet Connector

Heat Sink

Ready to Go
Bullet Connector

Channel 3

**Figure 3-18**    (*Above, right*) This ESC requires both battery and motor leads to be soldered—bullet connectors shown. (*Below, left*) This ESC just needs battery connection.

### Bullet Connectors

Soldering the bullet connectors is quite easy to do. A good high-wattage (40 watts or greater) soldering iron helps. As in all soldering tasks, the goal is to get the substrate metal hot enough to melt the solder and not just melt the solder on the iron and drip it onto the substrate metal. That said, there is a dirty little trick in which experienced solderers melt a bit of solder onto the tip of the iron to create a bridge of hot metal to the substrate. Be sure to check out the videos at brooklynaerodrome.com.

**ESC Side Bullet Connectors.**    Female bullet connectors are attached to the electronic speed controller (ESC) motor wires. Female bullet connectors have a wire side that is much shallower than the connector side—verify that you are soldering the right side by inserting the male connector before soldering. Steps include

1. Position helping hands (the alligator clips on a flexible joint) to hold the female connector.
2. Tin the speed control motor lead even if it is pretinned. This ensures that same temperature solders will be mating. If you don't know how to tin, check out our videos on the book website.
3. Stick the end of the soldering iron into the small hole in the bullet connector to heat the housing if it fits and the hole exists. Otherwise,

**Figure 3-19**    Bullet connector: filling the solder cup with solder.

apply the soldering iron to the side of the bullet connector, and use solder to create a heat-transfer blob to the connector.

4. Fill the connector with hot solder.

5. Plunge the motor lead into the connector while retaining heat for around 5 seconds to ensure that the tinned solder on the lead has melted and integrated with the connector, as shown in Figures 3-20 through 3-23.

6. Repeat for the other two connectors.

7. Apply heat-shrink tubing to the edge of the female connector as shown in the made-up connectors in Figure 3-18. You also can use a cigarette lighter or the side of the soldering iron to get the tubing to shrink. Note that heat-shrink tubing contracts to about 50 percent of its initial diameter. If the connection is 3 millimeters, then 6 millimeters of heat-shrink tubing should work, but be a bit conservative and go with 5 millimeters if possible.

**FIGURE 3-20**   Bullet connector: about to dunk tinned motor lead.

**FIGURE 3-21**   Bullet connector: dunking tinned speed control lead into pool of molten solder.

**Figure 3-22**   Bullet connector: continuing to apply heat for a few seconds to ensure a good joint.

**Figure 3-23**   Bullet connector: allowing the solder to cool.

**Motor Bullet Connectors.**    Most motors come with male bullet connectors soldered on already. If not, follow the same steps as for the female connectors with some critical differences.

1. Keep solder out of the male part. The connector will never fit with extra solder there.
2. Heat-shrink tubing should stop at the shoulder of the male connector, as seen clearly in Figure 3-18.

## Battery Connector

The solder-based connectors are fussy and are not as tolerant of soldering abuse as the bullet connectors because they have plastic bits that get hot and deform when they are being soldered. On deformation, they won't fit the battery. There are three strategies for this:

1. Don't overheat the plastic.
2. Get spare connectors when you order.
3. You can sometimes get male-female paired units. In this case, use the "battery" sided connector as a form-preserving shape to keep the plastic from deforming when it gets hot. This is what we do at Brooklyn Aerodrome with classes and novice solderers. *Do not use* a battery connector connected to a battery because you will likely short out the LiPo battery with inadvertent contact, and you may destroy the soldering iron.

The universe of battery connectors fans out quite a bit. The major classes of connectors are addressed below, and most likely what you have will fit one of these approaches.

**XT60.**    The XT60 is becoming a standard connector and is not difficult to solder. The plastic can get soft with excess heat, though. Figure 3-24 shows a deformed connector next to a normal one. The steps for soldering the connectors are similar to those for bullet connectors but with less plunging. Steps include

1. Use helping hands or Vise-Grips or a friend with pliers to hold the connector. If you have a spare female connector, use it to hold the male connector. It will protect against the contacts getting out of position from the plastic getting soft. *Do not use* the actual battery connector. You are very likely to short out the battery.
2. Almost all connectors have polarity protection, and the XT60 is no exception. The red wire goes on with the positive (+) marking and the black wire with the negative (−) marking. Verify with the battery that black connects with black and the red connects with red.

**Figure 3-24**  XT60. Top right pin has moved up and to the left from excess heat. Connector no longer fits the female end.

3. Tin the speed control wires.
4. Put ¾ inch of 4-millimeter- or ³⁄₁₆-inch-diameter heat-shrink tubing on the wires.
5. Align the correct wire with helping hands with one edge of the wire against the tube side of the connector, as shown in Figure 3-25.
6. Place the tip of the soldering iron on top of the wire, and feed a bit of solder to get heat transfer working. Then feed the rest of the solder onto the gold-plated connector until the joint looks good and saturated and strong. This should not take more than 5 to 8 seconds or the plastic may get too soft. Wait for the whole thing to cool.
7. Repeat for the other wire—don't forget to add the heat shrink.
8. Apply heat to the heat-shrink tubing with a heat gun, cigarette lighter, or the tip of the soldering iron to insulate the connectors.

**Cheetah Connectors.**    Cheetah connectors (Figure 3-26) are nice to use, but they can be a significant challenge to solder. They solder on just like bullet connectors (see earlier) with a requirement that little or no solder get on the outside of the connector or it may interfere with the ability to slide the plastic

**FIGURE 3-25** XT60 male connector being soldered, with a spare female connector being used to keep pins aligned. Do not use a live battery connector!

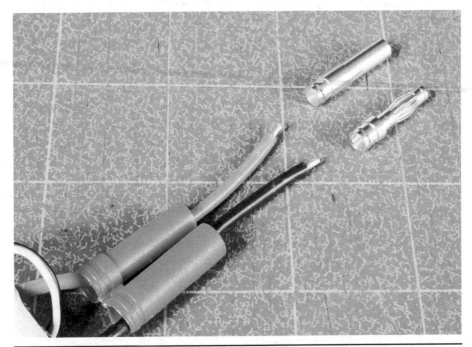

**FIGURE 3-26** Cheetah connectors. Put the insulating sleeve on before soldering, and be careful not to get solder on the outside of the connector.

housing on the outside and have the connector seat properly. Instructions include

1. Slide the plastic polarity housing onto the wires in appropriate orientation for mating with the battery.
2. Select the positive (red/+) lead, and set up soldering for the female connector (or the same as what goes on the speed control side of the bullet connectors).
3. Double-check that polarities are correct.
4. Solder as per the bullet connector instructions given earlier with extra care to keep solder away from the outside of the connector. A little is probably tolerable; a lot and the housing will not seat properly.
5. Solder the male connector.
6. Get a heat source such as a heat gun or cigarette lighter, and warm the plastic housing considerably without catching it on fire. With gloves, pull the housing down onto the connectors. You should feel a click as they seat, and they will not come out. If they don't seat, try heating the plastic more or heating the gold connector. Keep trying; you may have to pull pretty hard. Do one wire at a time. Be patient. Be firm.

## Power Up the Airplane

The time has come to bring the deck to life. On a stormy night, raise the lightning arrestor in your mad-scientist lair and do the following:

1. Dig out the receiver that came with your transmitter.
2. Plug the speed control receiver lead into channel 3. Black or brown is the ground or negative (–) lead; red positive (+) and signal white or yellow.
3. Plug the pilot's right (facing-front) servo into channel 1 and the left servo into channel 2 while paying attention to getting positive/ negative/signal orientation correct.
4. Verify that all receiver connections are appropriate for your receiver— remember that black or brown indicates negative, red positive, and white or yellow signal. Correct, if necessary, with consultation of your radio manual.
5. Remove the propeller from the propeller mount or prop saver if attached. The motor may start spinning, so removing the prop avoids possible damage/injury.
6. Connect the motor bullet connectors. There is no polarity to worry about; connect any motor connector to any of the speed control outputs. If the motor turns in the wrong direction, it can be reversed by swapping any two connectors.
7. Verify that the battery plug will connect with correct polarity to the speed control—black goes to black, and red goes to red. There *is* polarity

to worry about, so do not plug anything in yet. Check it again—if the polarity is wrong, the speed control and likely the battery will be destroyed if they are connected the wrong way.

Get your transmitter out, and add batteries, if needed. The radio startup sequence always should proceed as follows in normal operation:

1. The throttle control (left control) should be down or at zero.
2. Power up the transmitter. The transmitter should be first on and last off. This avoids the receiver picking up errant signals from the ether and starting the motor or moving servos to damaging extremes.
3. Power up the airframe by attaching the battery to the speed control. There is no switch on most speed controls.
4. The servos should come to life, and the right stick should move the servos. Since the arms are not on the output shafts, place a finger on each to be sure that they are moving. There may be a delay of a few seconds, and the speed controller may chirp a few times or wildly. If the aircraft does not come to life, then check that all polarities are correct at the receiver. If it still does not work, then investigate as detailed in Chapter 6 on troubleshooting.
5. You now have live electronics. Advance the throttle a third to verify that the motor turns for a few seconds. If it doesn't turn, then check your speed control manual for what is expected to happen before it will turn the motor.
6. Verify that the motor turns clockwise looking from back to front— orientation matters. Put your finger gently on the bell housing as the motor is moving at a fairly low speed to check. If it is turning counterclockwise, then power down the plane, swap any two of the three motor leads, power up, and it should be turning clockwise now.
7. Say something like, "It's alive . . . it's aliiiiive! My plans for world domination have begun!" Keep it short, but savor the moment. You still have lots to do.

Do not run the motor at high speeds or for long amounts of time.

## Powering Down the Airplane

The power-down sequence is the reverse of powering up:

1. Disconnect the battery from the speed control, taking care not to damage anything if the connector is sticky.
2. Turn off the transmitter.

## Securing Parts to the Deck

At this point you have a live aircraft, but the various components need to be protected from the high forces of crashes and typical flying. The deck needs to have all of its components secured. Make sure that there are no tight wires that might sheer in a crash or wires so loose that they can be drawn back into the propeller. Remember, most crash energy forces components forward, so defend against that. Steps include

1. Add the hook side of the hook, and loop tape to the end of the deck. In cold weather, Velcro tape adhesive can release, so adding a few zip ties in the middle and at the ends will help to secure the hook tape.
2. Attach the loop side of the tape to the battery, and attach it to the end of the deck. The battery is now placed correctly, and the rest of the components need to be secured while allowing all relevant parts to be connected.
3. Secure the receiver around the motor-mount area. A good, protected spot is in the elbow of the angle stock, but off to the side is fine, too. Details are up to the builder, but top-plug receivers provide a natural bump to put a zip tie across to resist being flung forward on a crash as shown in Figure 3-27. End-plug receivers can be a bit more difficult to secure, but a lengthwise zip tie can handle it nicely.
4. If appropriate, secure the antenna of the receiver so that it won't get damaged.
5. Servo wires tend to come loose on crashes and then get sucked into the propeller. The best approach is to loop any excess servo wire around a zip tie as shown in Figure 3-27 or try the Make Labs trick below. Tape is a bad idea for securing wires if crashing is expected.
6. The folks at Make Labs figured out a nifty way to route wire. Cut a slit next to the servo, run the wire under the deck, and cut another slit next to the receiver for a very clean installation. Thanks to Dan from *Make Magazine*, who came up with this.
7. The speed control has to be placed where it can connect with the receiver and allow the battery pack a range of placements to adjust the airplane's center of gravity. Be sure to put a zip tie fore to aft to resist crash forces and keep the smooth-sided metal heat sink up in the airflow for cooling.

## Conclusion

The deck is done, as is most of the work. The deck is the resilient part of the Flack that will live through many wings. In Chapter 4 we build the airframe.

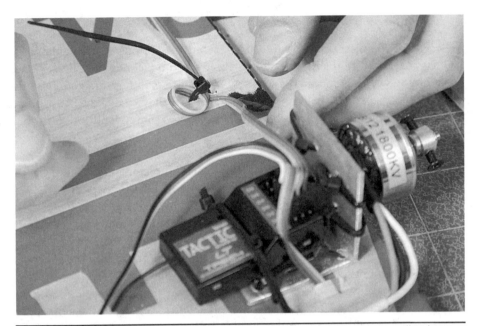

**FIGURE 3-27**    Securing the deck. Looping extralong wires around zip ties works well to keep them out of harm's way.

# The Airframe of the Flack

If the deck is the soul of the Flack, then the airframe is the first of many bodies. While the airframes are remarkably tough, they eventually get soft after around 30 or so hard crashes. When that happens, a new airframe can be created and the deck transferred in the Brooklyn Aerodrome's version of reincarnation. We have decks that have been reincarnated at least 10 times. Whether there is a metaphysical lesson to be learned here is up to the reader.

This chapter covers how to create the airframe and attach the deck to it. The airframe consists of the wing, control surfaces, and stabilizers. Just a few steps require precision: (1) cutting the elevon hinge and attaching it, (2) getting correct trim on the control surfaces, and (3) getting the center of gravity (CG) right. This is also a good time to spend some time on a flight simulator. Enough said. Let's get building! Behold the object of your efforts—the Flack—in Figure 4-1.

## The Airframe

Be sure to check brooklynaerodrome.com for advancements on airframe design. However, any advancements will be close to what we do here. The construction material of choice is Dow fan-fold, but all materials will employ the same basic steps.

If the foam was shipped, it may have been cut to keep shipping costs reasonable. For example, Brooklyn Aerodrome kits need to have tape applied as shown to re-create the airframe pictured in Figure 4-2. If not, then skip to tracing the plans onto the foam/cardboard/wing material.

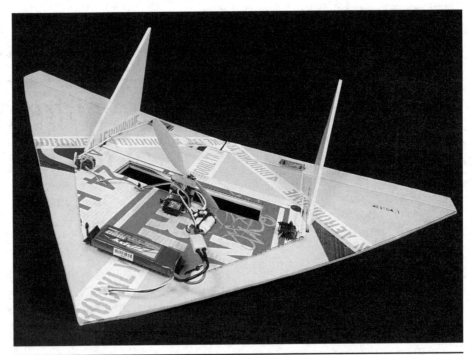

**FIGURE 4-1**    The Flack ("flying" + "hack").

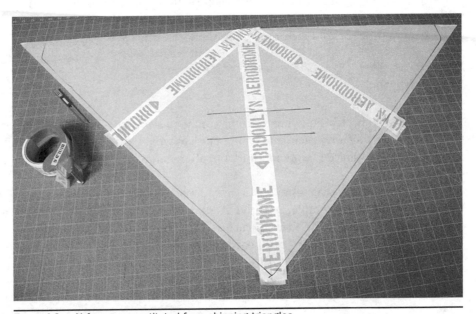

**FIGURE 4-2**    Airframe reconstituted from shipping triangles.

The steps to create a raw airframe from the kit include

1. Puzzle together the pieces of foam as shown in Figure 4-2.
2. Tape the pieces together using a tension-maintaining technique. This works as follows:
   a. Align the foam edges.
   b. Get a 2-inch length or so of packing tape.
   c. Hold up one side of the joint, and apply the tape to the angle at one end of the joint.
   d. Lay the foam flat, and the tape joint should be very tight.
   e. Do the same for the other end of the joint, as shown in Figure 4-3.
   f. Lay the foam flat, and apply tape along the length of the seam.
   g. Flip the foam, and tape the length of the other side.

This technique for rejoining foam has been used for years at the Brooklyn Aerodrome. It is how we get as many airplanes out of a block of foam as possible.

Now that there is a big enough piece of foam to trace the plans, let's cut a wing.

**Figure 4-3**  Taping foam triangles to create the wing. Hold the foam at an angle when applying the tape to ensure a tight seam.

1. Trace the plans from Figure 4-4 onto the foam. If the foam is a full 24- × 48-inch sheet, then place the triangle in the middle to allow creation of another airframe from the two remaining triangles.
2. Cut the outside shape from the foam using a sharp blade. Do not cut the elevons out yet. A dull blade will tear at the bottom side of the cut. You can use a ruler to keep the cuts straight, but the edges turn out just as well with long, smooth hand cuts.
3. Draw in the elevon hinge line on the surface you want to have on the top 2 inches back from the rear. It is recommend that you write "Top" on the wing.
4. Cut all the way through the foam at a consistent angle (around 30 degrees) along the entire length of the elevon, as shown in Figures 4-5 and 4-6. This cut needs to be straight and is easier to do with a ruler. It is worth practicing making this cut and hinge on some scrap to understand what is going on.
5. Flip the wing over so that the acute angles align on top, as shown in Figure 4-7.
6. Use 2-inch sections of tape to hinge the touching angles at the end, middle, and other end of the elevon, making sure that there is no space between the angles. Then tape the entire length of the hinge—several 1-foot sections are easier to work with than doing it all in one 42-inch

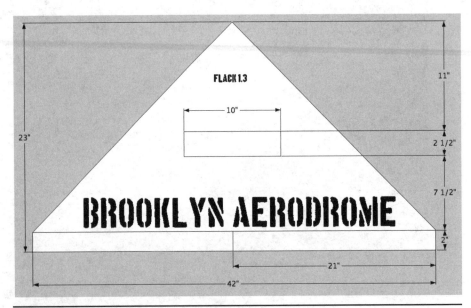

FLACK 1.3

11"

10"

23"

2 1/2"

7 1/2"

**BROOKLYN AERODROME**

2"

21"

42"

**Figure 4-4**   The Flack 1.3.

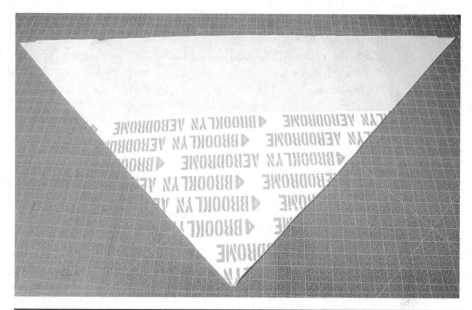

**FIGURE 4-8**    Nose reinforcement for beginners.

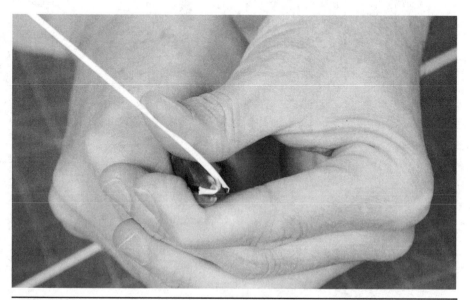

**FIGURE 4-9**    Control rods. Bend a piece of coat hanger around the end of needle-nose pliers.

3. Measure 11 inches from the back of the U, and mark on both control rods at the same time.

4. Place the pliers at the mark, and bend the coat hanger 90 degrees to make an L. Keep the bend in the same plane as the U bend. Note that the bend goes in the opposite direction from the U bend, as shown in Figure 4-9. Repeat for the other control rod (Figure 4-10).

5. Rarely does the L bend actually end up in the same plane as the U bend. Place the U bend flat on a level surface to test that the L bend also lies flat. In Figure 4-11, the L and U bends are not aligned. To fix, hold the U bend in the pliers, and twist the L bend so that the bends lie in the same plane (Figure 4-12).

6. Verify that the control rods are within $\frac{1}{16}$ inch of each other. If one is significantly shorter, then use the technique for establishing reflex trim discussed below to shorten the longer one or just try again.

7. Trim the L bend to 2 inches.

**Figure 4-10**   A pair of matched control arms.

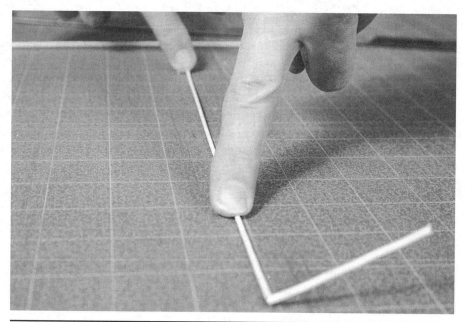

**FIGURE 4-11**    L bend out of the plane of the U bend.

**FIGURE 4-12**    L bend in the same plane as the U bend.

### Attaching the Control Rods to the Elevons

I use Coroplast material for attaching the control rods to the elevons. The elevons are simple but do require some precision in placement. The goal is to match the left and right control-horn placements. This will ensure that up and down commands will not introduce roll because the elevons move different amounts for the same inputs. The end result of these steps for each control horn is shown in Figures 4-13 and 4-14. Note that the elevon horns are very close to being the same distance from the tape hinge line—the control-rod hinge point is what matters, not the Coroplast.

Steps for the build include

1. Place the tape hinge side of the airframe down; the bevel is up.
2. Make a 2- × 2-inch piece of Coroplast scrap material, and cut it in half across the grain of the flutes.
3. Carefully use the end of the L bend to make a hole through the flutes of the Coroplast about $\frac{1}{16}$ inch from the edge. Try to keep the coat hanger parallel with the edge of the Coroplast. Wiggling helps, but be careful to not stab your hand when the wire comes out the other side (see Figure 4-15).

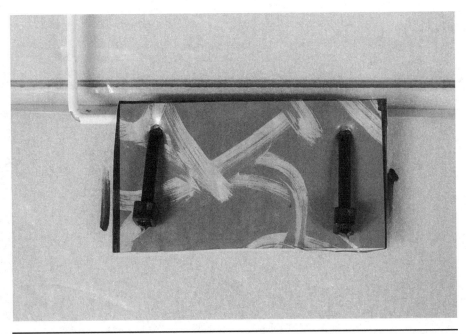

**FIGURE 4-13**  Left elevon hinge placement.

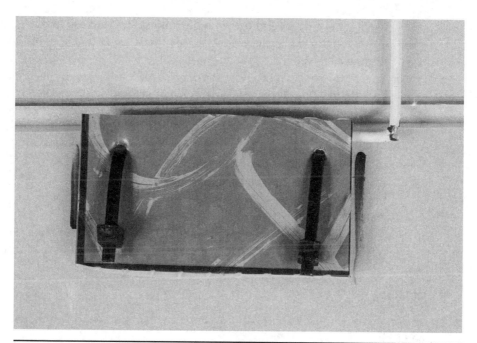

**FIGURE 4-14**    Right elevon hinge placement. Note how the coat hanger hinge point is the same distance from the V at the bottom of the bevel as on the left side.

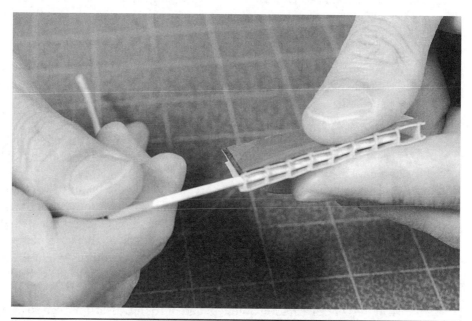

**FIGURE 4-15**    Creating the hinge by forcing the coat hanger through the flutes.

4. Placing the deck over the prop hole, follow straight back from the servo output shafts to the elevon. Mark this location—it is where the control horn needs to be left/right on the elevon.

5. Trial fit the control horn with the control rod parallel with the hinge line and about $\frac{1}{16}$ inch back from the bevel. The important consideration here is that the hinge line be parallel with the hinge point of the control rod, as shown in Figures 4-13 and 4-14.

6. Once you know the location of the control horn, apply double-stick tape to the elevon side of the Coroplast, and attach to the elevon.

7. Repeat the preceding steps with the other elevon/control rod, paying particular attention to mirroring the distance of the elevon hinge to control-rod hinge of the first elevon.

8. Secure both Coroplast control horns with zip ties to ensure that they do not come loose. If one control horn comes loose, it is a guaranteed crash.

## Attaching the Deck

We are almost done. Next, we do the fundamental alignment step that sets up the Flack.

### Drill and Attach the Servo Arms

The servo arms need to be drilled out to accommodate the diameter of the control rods. You need two drilled arms. Making an extra one is not a bad idea for a crash kit. A crash kit is a collection of spare parts and tools for taking to the flying field—see Chapter 6 for more information. Steps include

1. Select a drill bit that is the same or slightly smaller than the size of the coat hanger wire. In a tight spot, you can use a bit of coat hanger to drill the hole.

2. If you are a beginner, drill about $\frac{1}{4}$ inch out from the spline hole. This is usually the second hole out. If you know how to fly, you can drill farther out on the arm to get increased control throws or drill both and decide on the field.

3. The servo horns in Figure 4-16 are drilled for both beginner (closer to the servo-shaft hole) and advanced control throws (farthest from the servo-shaft hole). The servo-horn retaining screw is on the right. The two servo retaining screws on the left are not used, but keep them around because in a pinch they can be used to retain the servo horns when there has been slight violence to the servo output shaft. Double-sided horns can be used as well by cutting the extra arm off as shown.

4. If the control arms are double-sided or round, just trim them so as not to interfere with the deck.

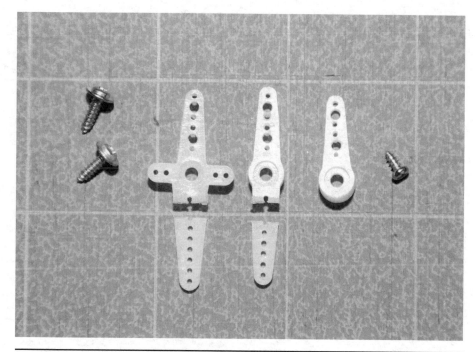

**FIGURE 4-16**   Drilled and trimmed servo horns—only two needed.

5. Insert the control rods into the holes. Make sure that the arm can move freely—the U can be too narrow and bind, preventing free movement. Adjust as necessary.

6. Make sure that there is no slop in the connection between the servo horn and the U bend. During a test build, one builder ran the correctly sized drill bit backwards, which made the hole too big, resulting in a loose fit.

7. Attach the servo arm to the splined shaft of the servo. Do not put the servo-horn retaining screw in yet.

8. Adjust the servos by moving the arms gently with your hands. Be sure that the servo arms are at 90 degrees to the deck, as shown in Figure 4-17. Be gentle in moving the servos. Do not do this with servos powered.

### Aligning and Attaching the Deck

This is the moment when the whole airplane comes together. Check everything twice, make sure that the servos have not moved from 90 degrees, and be patient. The steps include

1. Put some weight on the back of the airframe between the two control horns and the rear of the wing to keep the elevons flat to the surface.

**Figure 4-17**  Servo arm at 90 degrees to deck surface.

2. Adjust the deck to be centered left-right on prop hole. *The important thing is that the servos be at 90 degrees and the rear of the wing and elevons be flat.* If the deck prop hole and wing prop hole do not match exactly, that is okay. If the prop is a little skewed in the prop hole, that is also okay. This is not a crucial dimension, believe it or not.

3. Flip the deck up, and apply double-stick tape to the underside (Figure 4-18).

4. Flip the deck back, and place on pens or pencils so that the tape does not make contact with the airframe. The pens or pencils allow for a final alignment check.

5. There are two things to check—this is *important*. The elevons and rear of the wing should be flat on the surface, and both servos should be at 90 degrees to the deck. Check again. If the prop hole of the deck doesn't quite match the prop hole on the wing, don't worry about it. You can cut out more foam to give extra clearance.

6. Remove the pens or pencils, and press the deck into place on the wing gently, starting from front to back. Starting at the back can result in the elevons being raised. Check the elevon trim—if either is more than ¼ inch up from the tabletop, then gently pry up the deck and redo the placement. If the elevons are a little down, this is not a problem because there are ways to make the control horns shorter. Once satisfied, press down firmly.

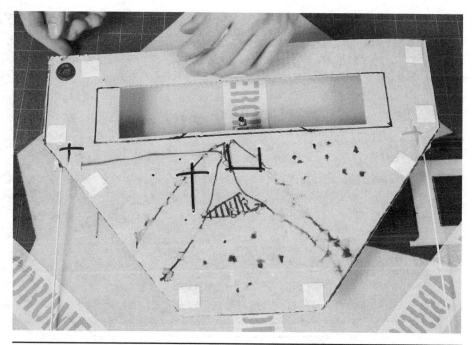

**Figure 4-18**   Deck with double-stick tape attached. Servos are still attached to the control rods.

7. Insert zip ties at the front corners by the battery pack and at the rear edge of the deck. You also can add zip ties around the servos if you think the tape might not hold the deck well enough.

## Centering Controls

The next step is to center the servos, attach them, and verify control movements. Steps include

1. Establish elevon mixing with the transmitter. Consult your user manual.
2. Cut ¼ inch between the elevons in the center so that they move freely.
3. Move the aileron and elevator trim tabs to the neutral position on the right-hand stick, as shown in Figure 4-19. Neutral is in the middle of the tab's range. If you have digital trims (rocker switches that control trim), consult your radio's manual for how to know that you have neutral trim.
4. Remove the servo output horns from the servo shafts, but keep them on the control rods.
5. Remove the prop.

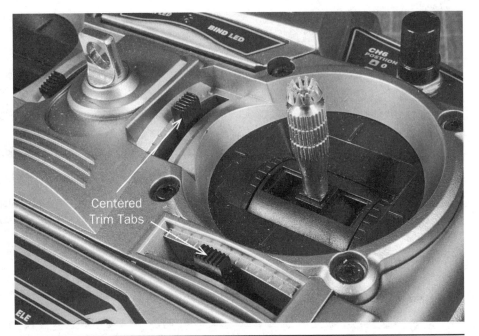

FIGURE **4-19**    Trim tabs at center on right stick.

6. Power up the transmitter with the throttle down.
7. Power up the airframe.
8. Attach the output horns to the closest spline to being vertical. If one arm is slightly forward, then let the other arm be vertical or slightly forward as well. Try to match the positions of the servo arms. Do *not* force the arms when the servos are powered—they will resist, and you may damage the gears.
9. Power down the airframe.
10. Apply the servo-horn retaining screws with care. You can strip the servo gears if too much force is used. Snug is good.

### Establish Elevon Trim

1. Power up the transmitter and Flack.
2. Verify that the controls work as expected. Figures 4-21 through 4-24 show control stick and corresponding elevon deflections. Check this twice. You may have to revisit your transmitter manual to reverse channels or mixing functions.
3. Use only the trim tabs (not the control stick) to create the ¼-inch up reflex with both elevons, as shown in Figure 4-20. The foam is ¼ inch

**FIGURE 4-20**    The ¼-inch up reflex.

**FIGURE 4-21**    Neutral elevator, neutral aileron.

**FIGURE 4-22**   Down elevator, neutral aileron.

**FIGURE 4-23**   Up elevator, neutral aileron.

**Figure 4-24**    Neutral elevator, right aileron. Left aileron is reverse.

thick, so use that as a measure if you have it. You may need to make adjustments to aileron and elevon trim tabs.

4. If the trim tabs have insufficient throw to establish correct trim, then the control rods will need to be physically adjusted. See the section on physically adjusting trim below. It is a good idea to have the correct flight trim done physically anyway with the transmitter trim tabs in a neutral position.
5. Power down the Flack and transmitter.

### Physically Adjusting Trim

The Flack is built in a fairly loose fashion, so it is not unexpected that the correct trim of the plane is beyond the range of the trim tabs on the transmitter. In general, it is better not to use trim tabs at all to get flight trim for a few reasons:

1. Trim is meant to be adjustable for different flight modes. Flying fast will use less up trim; slow will need more; etc.
2. Trim is useful when cargo is being carried. If you need a bit more up trim because of the camera mounted on the nose, then it is annoying not to have any extra up trim left from just getting the plane flying normally.
3. If you fly more than one model with the transmitter, then you will be able to swap models without worrying about what the relevant trim is supposed to be with the cheaper radios. More sophisticated radios store trims with individual models in memory.

The process of physically adjusting trim is as follows:

1. Be sure that the transmitter mixing is correct; otherwise, your physical corrections may not be working from correct centers. On cheaper radios, trim centers may change between mixing settings.
2. Center trim tabs before making adjustments.
3. To make a control rod shorter, consult Figures 4-25 through 4-27. Practice on some spare coat hanger material before doing it to the actual control rods.
4. There is no easy way to make control rods longer, but you can remove the retaining screw and slip the servo control horns one indent of the spline shaft back to get down trim. Both servo arms should be matched, however, if this is done. Alternatively, just make a new control rod that is a bit longer.

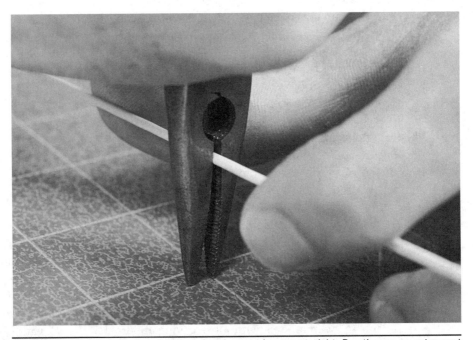

**Figure 4-25** Twist clockwise while keeping the coat hanger straight. Practice on an extra coat hanger.

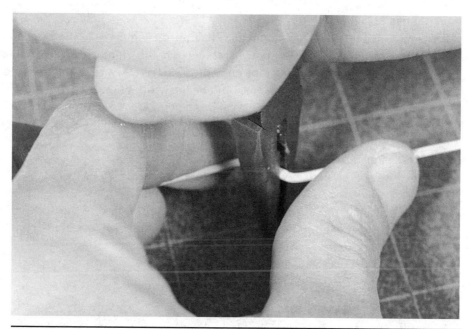

**FIGURE 4-26**    Grip with pliers and fingers. This takes hand strength.

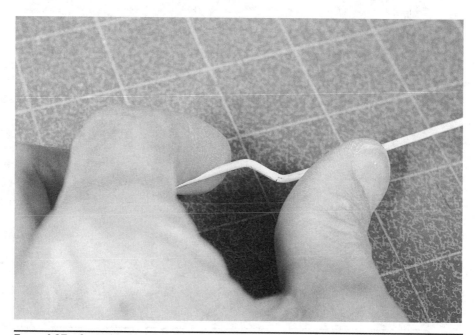

**FIGURE 4-27**    Control rod is now shorter.

### Cutting Stabilizers

Draw out two stabilizers from the diagram in Figure 4-28. Both tabs at the bottom need a ¼-inch cutout, as shown on the right. The tabs will need reinforcement with packing tape on both sides. It is easiest to apply the tape first and cut after. The ¼-inch notch should be the width of the airframe material, as shown below. If you have fiber tape, it is a good idea to reinforce the stabilizer tabs.

### Cutting Stabilizer Slots

Next, we cut the slots in which to insert the stabilizers. Steps include

1. On the top side of the plane, put the stabilizer next to the deck and with its rear edge at the bevel of the hinge. The stabilizer should be next to the deck. Refer back to Figure 4-1 for their placement.
2. Move the stab ½ inch forward.

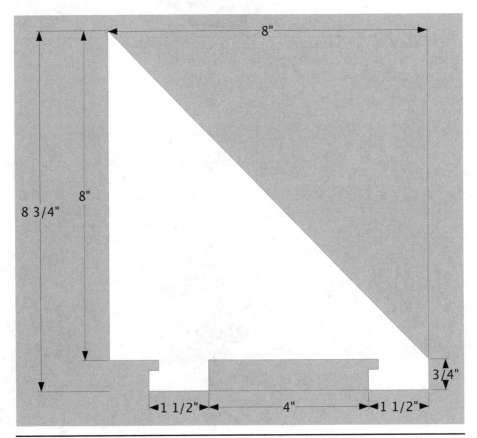

**FIGURE 4-28**  Stabilizer template.

3. Mark the stabilizer tabs with a pen. The idea is to have the stabilizers be nearly flush with the hinge line once they are inserted and pulled back.

4. Cut out the slots, being careful not to make the slots too wide—there is no problem if they are a bit too long.

5. Don't make the stabilizer slots so tight that it is hard to remove the stabilizers. When transporting the plane, you should remove the stabilizers to keep them from getting damaged.

Test insert a stabilizer without forcing the foam to deform. If the slot needs to be a little bigger, then make the adjustment. When finished, the rear of the stabilizer on the top side should be near or at the hinge line.

## Final Cleanup

The airplane is done.

1. Remove extraneous zip tie ends. Take care not to cut wires by mistake.

2. Attach the prop in the correct direction (raised writing toward the front).

3. Advance the throttle a third, and verify that the motor spins in the right direction (clockwise looking from back to front unless you have a special prop).

4. Grasp the plane firmly (be careful here), and apply full throttle for a few seconds. There should be a lot of thrust—papers blown off the desk—about a pound of force at full throttle.

5. Write your name, phone number, Academy of Model Aeronautics (AMA) number, e-mail address, and any reward you are willing to pay for return on the airframe.

6. Draw a line $10\frac{1}{2}$ inches from the nose the width of the prop hole. This is your center of gravity (CG).

7. Verify the CG. The CG is the point at which the airplane is balanced. The most important dimension is fore/aft, which can be tested by finding the balance point of the airplane with two fingers as shown in Figure 4-29—see Chapter 10 for more on what is going on. Adjust the CG by moving the battery fore or aft, and if this is insufficient, add nose or tail weight to achieve the CG. If you are building with heavier materials, expect to need to add nose weight; if you are building with lighter materials, then expect to add tail weight.

8. Verify that you have appropriate flight trim, which is both elevons up $\frac{1}{4}$ inch.

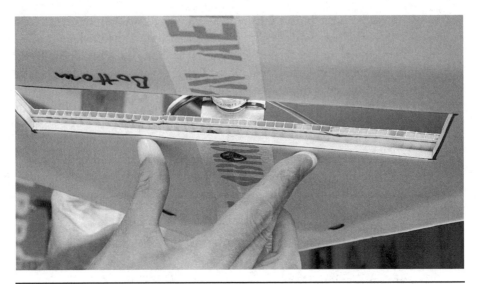

**FIGURE 4-29**   Verifying CG by balancing the aircraft on your fingers. Do this indoors.

## Conclusion

Now you are ready to fly. Chapter 5 takes you through Brooklyn Aerodrome's fun and proven approach to flight instruction. You are going to crash a lot—and have a great time doing it.

# CHAPTER 5

# Learning to Fly
# One Crash at a Time:
# The Splinter Method

Learning to fly from a book ought to be a punch line for a joke, but it works. Don't skip any steps, and it will go fine (Figure 5-1).

Our approach to learning to fly at the Brooklyn Aerodrome involves a lot of crashing. This is not how most people learn to fly. At the Brooklyn Aerodrome we call what we do the *Splinter Method*. Mark "Splinter" Harder learned to fly at night in a windy, tree-lined park in the middle of Brooklyn. He was going to crash a lot, and he did. This approach is named after him for surviving it and providing crucial feedback as to how to optimize the experience.

We have kept the Splinter method even when conditions don't mandate it. Why? First, it is a fun way to learn that teaches quickly what works and what does not from the pilot's perspective. Second, it also provides immediate independence for the pilot, and third, it quickly gets the "worst thing that can happen" out of the way, which reduces anxiety. Besides, all beginners crash, so you might as well make it a feature of the program.

## Safety

The airplane you just built is dangerous. It is meant to be no more dangerous than a well-hit softball, but this has not been verified or tested, nor is it guaranteed. Even though it is made of foam, the airplane has cut people, it has caught fire, and it has delivered some bruises. The pilot, even a beginner, is responsible for damage caused by his or her aircraft. Join the Academy of Model Aeronautics (AMA) for insurance protection that will cover damage you cause to others as an act of responsibility—see Chapter 1. Below are some points on how to keep safe and have fun.

**FIGURE 5-1**  Bethany first-day flying, left-hand launch.

## The Cone of Crashing Potential

In almost any flying spot, there are objects to avoid hitting—at the very least, the pilot, but other to-be-avoided objects include trees, people, animals, water, and property. The *cone of crashing potential* (CCP) is how far an out-of-control airplane will go with no power plus some margin of error. Figure 5-2 shows a projected CCP in a typically crowded park environment. The size of the CCP depends on air speed, altitude, wind, and the consequences of hitting objects. The idea behind the CCP is that if a to-be-avoided object is in that cone, the beginner pilot will immediately cut throttle and attempt to land the plane. This has some very desirable properties:

1. With throttle cut, the top speed of the airplane is minimized, and damage to the plane or what the plane hits is greatly lessened.
2. The CCP reinforces situational awareness for the pilot, which is key to successful flying skills. A spotter is very helpful where there are a lot of obstacles to be aware of.
3. The plane without power is more docile, easier to right, and less likely to overreact to inputs. When the pilot is more skilled, he or she is better able to know when power should be applied or cut in troublesome situations.

**FIGURE 5-2**    The cone of crashing potential visualized in a Brooklyn park.

## Trim

A well-built Flack with the trim shown in Chapter 4 (¼-inch up elevon) and the proper center of gravity (CG; 10½ inches back from the nose) should fly straight. During test builds for this book, Flacks flew straight or very close to straight off the building board. But things may change as a result of crashes or gear being added, so this section addresses trim and how to adjust it. For a beginner, the trim of the aircraft is very difficult to adjust because beginners don't know yet how airplanes should fly. A well-trimmed Flack flies as follows:

1. The Flack should be able to complete the glide task (task 1) while landing mostly flat without roll or pitching up or down.
2. Depending on the motor, the Flack launched at one-half to two-thirds throttle flies smartly out of one's hand (look for launch technique videos at brooklynaerodrome.com).
3. If the plane has been thrown with some skew or pitch or too slowly, corrective action may be required, so the plane needs to have any launch funkiness sorted out before assessing trim.
4. Once the plane is flying level in both pitch and roll, it should fly without control inputs in a straight path with a slight climb. The degree of climb can be controlled by throttle alone. In calm air, the plane should be able to fly 100 feet or more before corrective inputs are needed. In wind, this may or may not be possible.

A good place to learn to trim a plane is in a flight simulator. The plane should be trimmed to fly level in pitch and roll with no inputs from the elevator and aileron stick at two-thirds throttle. Try to fly at the lowest throttle setting you can in the simulator to keep the airplane from being overreactive. On launching, the Flack usually needs some up elevator that is no longer required once it is flying.

### Adjusting Trim

If the plane is rolling right, then the aileron trim tab should be adjusted a few clicks to the left. If the plane is rolling left, a few clicks of right trim are needed. If the plane is pitching up, then some down trim is needed, and the reverse is true if the plane is pitching down. The transmitter trim tabs are used to make permanent changes to the center of the control stick. On digital trims, there is a switch, and each actuation of this switch acts like a click. It often happens that the trim limits of the transmitter are reached before the plane will fly level. In such cases, the control rods need to be readjusted to achieve the last best flying trim on the control surfaces and the trim tabs recentered.

Out-of-trim airplanes are harder to fly. If you can enlist the help of a more experienced pilot to trim your airplane, you should consider doing so. If you can have someone else launch the plane, then do that. See launch instructions below. If you are flying by yourself, do your best to launch cleanly, and see what the airplane is doing.

## Flight Simulator Setup and Training

Plenty of paid and free flight simulators are available for Linux, Windows, and Macintosh computers. See Chapter 12 for how to set up various flight simulators, and practice a bunch before flying for real. USB transmitter controllers can be obtained for $20. A small investment will more than make up for the cost of replacement parts if you spend some quality time with a simulator. It should be noted that some transmitters can run the flight simulator as well. If the simulator has a wind adjustment and/or turbulence mode, be sure to experiment with them as well. Most simulators have a "chase" mode, where you are always looking at the plane from behind it. Use this feature sparingly because this is not how you really fly the plane. You need to learn how to react when the plane is coming at you (left and right inputs reversed).

## Incrementally Learning to Fly

In the absence of an instructor, learning to fly is a challenge, but it can be done. The process, preferably with a partner, starts by treating the Flack as a glider to verify basic trim and to learn to launch cleanly. The next step is to use the transmitter to help flair the landing, but still no throttle. Once the Flack can be regularly launched and landed without incident, the throttle is gradually increased over successive launches while the pilot learns to cut throttle before landing and to keep the wings and nose level. When there is sufficient power to keep the Flack airborne, flights are gradually increased from 50 to 100 feet at altitudes no greater than 20 feet. Once this is mastered, then it is time to learn to turn. The system works. It has been tested on beginners using only the text in this chapter with 100 percent success. Remember that the goal of your first flights is to get in the air and get back down without trauma, drama, or bad karma.

## Get a Helper/Partner/Friend to Help Out

Orville and Wilbur Wright had to learn to fly without any help, but crucial to their success was that they had each other. In the spirit of their efforts, I strongly recommend that you have someone help you in your learning because it allows you to focus on flying and not on launching, spotting, and other distractions. For this reason, Brooklyn Aerodrome's children's classes always involve pairs of children or a child and a parent. That said, our first three testers (Joseph, Andrew, and Rob) did it alone.

## Control Sensitivity

The Flack is a very responsive airplane, even with the beginner's set up with inner control-horn throws. No trainer airplane performs this way to our knowledge, so why do we do it? The real reason is that these design parameters were necessary to fly in typical gusty urban conditions. Hundreds of people have learned to fly with a Flack, so why mess with success? One guess as to our success is that beginners quickly see what the control input does to the plane and adjust more rapidly—particularly kids. The Flack will mostly fly itself, with minor adjustments constantly helping it along. If the student makes a mistake, the correction will immediately fix things or make them worse. This instant feedback gives the pilot a clear signal about what works and what doesn't work. Typical remote-control (RC) trainers will take a long time (a second or two) to respond to inputs because of the limited control authority and dialed-in stability. The feedback lag leads to the beginner trying all sorts of things that don't really make a difference until things get bad, and then there is insufficient control authority to correct a bad situation even when the pilot tries the inputs that would have worked to correct the problem.

If your transmitter has dual rates, then learn with high rates or maximal control throw. When you start flying faster, then you can use dual rates to limit responsiveness. The Flack is flying so slow for the initial sequence that you need the extra control authority to control the plane.

## Crashes

You are going to crash. Promise. Some strategies to deal with crashes include

1. For a beginner, there will be a recognizable moment when the airplane is out of control. An essential skill is to recognize loss of control, cut throttle, and get the wings and nose level.
2. Once a plane is down, verify that the throttle is cut. Otherwise, you risk burning out the speed control and motor. This is a hard habit to instill.
3. Diving crashes are the most damaging. Learn to get the nose up before hitting the ground, and the crashes will start to look like aggressive landings.
4. On a hard landing/crash, do not pick up the airplane until all the parts have been accounted for. Batteries go flying, and prop savers spin off. It can be very hard to find ejected parts, and this is made worse when the crash site has been walked away from.

## Task 0: Approach the Field Ready to Fly

Never go out to the flying field with anything less than a perfectly setup plane, charged batteries, and confidence that no further work is required on the Flack or radio. A classic mistake is to show up at the field with a radio that is not programmed properly—this rarely ends well. It is agony to spend 20 minutes in the field on a task that could be done in 3 minutes in the shop. Before heading out to the field to fly, power up the airframe in the studio, and do a complete preflight check inside to satisfy yourself that nothing has been overlooked. A range test should be conducted before the first flight. Your transmitter manual should tell you how to do this. Generally, there is a button that you can push that reduces transmitter power, and you test for control of the Flack at a reduced range. Then power down and go flying.

## Task 1: Launching without Power

The first launch of your beautiful new Flack will be as a glider and without RC. It is the best way to learn to launch without the added pressure of trying to control the aircraft. It also establishes the fact that the plane flies fine without your input at all. You will most likely begin to crinkle your plane up a bit with the uncontrolled and unpowered launches, but don't worry about it. These little dings are saving you from much bigger dings later. Try to do this over grass. If you have an assistant, he or she will take over the launching task in the instructions that follow. Steps for task 1 include

1. Verify the CG—10½ inches from the nose is ideal—it should be marked clearly on the bottom.

2. Attach the battery, and power up the transmitter and then the airframe.

3. Put the transmitter on a safe and dry spot on the ground.

4. Until you get it so that you can land the Flack flat or with just a slight dive:

   a. Grab the Flack just ahead of the prop hole. If you are launching for yourself, I suggest that you use your left hand—look at Figure 5-3. If you are right-handed, it may be more comfortable starting with your right hand, but transition to your left hand when you get the hang of it if you are learning alone. The reason is that the elevator and aileron should be available immediately on launch to correct for wind, skewed launches, etc.

   b. Verify that you have the correct trim of ¼ inch up elevon. The elevons should match. Do this every launch.

   c. Rotate the prop so that it is not going to get pinched on landing.

   d. Prepare to launch. If there is wind, face into it. See the wind section below.

   e. Smartly accelerate the plane sidearm, flat, level, and with enough force to have it land 5 to 10 feet out, as shown in Figures 5-4 through 5-7. Look at Figures 5-8 through 5-11 as well.

   f. When you can launch cleanly and land with just a slight nose in or flat, you are ready for the next step. If this takes more than 10 tries, then move to task 2 anyway.

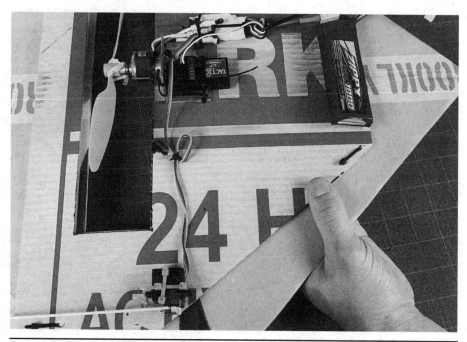

**Figure 5-3**   Proper launch grip.

**FIGURE 5-4**    Proper launch—no power or RC.

**FIGURE 5-5**    Plane is level, not thrown up or down.

**FIGURE 5-6**    Flack dives a bit. This would be handled with an up elevator on RC launch.

**FIGURE 5-7**    Plane landed 10 feet out.

## Task 2: RC on Launch to Landing, No Power

It can be useful to have someone hold the Flack in non-level attitudes and for the pilot to issue the corrective commands. The pilot should be behind the airplane initially.

1. If the nose is pointing up, apply down elevator until the nose is level.
2. If the nose is pointing down, apply up elevator until the nose is level.
3. If wing is rolled right, apply left aileron until level.
4. If wing is rolled left, apply right aileron until level.

When level flight is mastered, then the exercise can be used with the airplane's nose toward the pilot—this will reverse left/right roll commands.

The next step is to train your flight-deprived brain on how to flare the Flack on landing and correct any roll issues from launch. *Important:* A very common mistake when retrieving your airplane is to accidentally brush the throttle control with your coat, hand, etc. and start the motor turning when you are not expecting it. Nasty cuts have occurred. Always put your radio down on a dry spot with the throttle down before heading over to pick up your airplane. The steps are same as task 1, but now the transmitter is held by the pilot and is driving the servos. No power to the motor yet, though. Proceed as follows:

1. Grip the Flack as in task 1.
2. Verify that you have servo movement.
3. Say "Up," put in full up elevator, and look at the elevons to confirm that they are going up. This will help to train your brain about "airplane up" versus "video game up."
4. Launch as before, with the goal being to land 10 feet away (Figure 5-8).
5. Use the controls to keep the Flack level in pitch and roll (Figures 5-9 and 5-10).
6. Be prepared to flare the landing (Figure 5-11). Flare is achieved by raising the nose about 5 degrees at about 1 foot of altitude to soften the landing by reducing airspeed and getting the nose up.
7. Put down the transmitter and retrieve the Flack.
8. Repeat this process until the launches and landings are smooth and corrections are working. If you are a serious video gamer, you may have to spend quite a bit of time unlearning the notion of up.

**FIGURE 5-8**    Low-power sidearm launch—arm straight, plane level.

**FIGURE 5-9**    Slight left roll on launch.

**FIGURE 5-10**    Roll still not corrected but not bad.

**FIGURE 5-11**    Landed 15 feet away.

### Task 3: RC on Launch and Landing, Low Power, Cut Throttle

If you skipped either task 1 or task 2, go back and do them—seriously, they make a big difference.

This sequence will add to your launching and landing skills the ability to cut throttle before landing/crashing. Learning to cut the throttle is essential to the longevity of your airplane because it saves propellers and prevents the motor from being stalled (prop held by the ground with power being applied), which may burn out both the motor and/or the speed controller.

1. Grip the Flack as in tasks 1 and 2.
2. Verify that you have servo movement.
3. Say "Up," and visually confirm that you have up elevon movement.
4. Advance the throttle until the propeller is just moving—the lowest throttle setting that moves the propeller smoothly. See Figure 5-12 for how to use your chin if you are alone.
5. Launch with the intention of landing 10 feet out.
6. Cut the throttle before you flare the landing. Again, it is vital that the motor not be stalled while power is applied. This will burn out the motor and/or speed controller.
7. Put down the transmitter and retrieve the airplane.
8. Repeat at this level of power until you are comfortable cutting the throttle prior to landing the airplane.

### Task 4: RC on Launch and Landing, Increasing Power

The Flack is overpowered. Depending on motor/prop/freshness of battery, the plane will fly at between one-half and two-thirds throttle. When learning, use just enough throttle to keep the plane in the air. The reasons for this include less energy on crashes, which means fewer repairs, and a less responsive plane and slower flight speed, which give the pilot more time to think. All those excessive control throws work really well at these low speeds. In the next steps you will work your way up to flying 50 feet straight and landing.

1. Grip the Flack.
2. Verify that you have servo movement.
3. Say "Up," and visually confirm that you have up elevon movement.
4. Advance the throttle "a little bit more" than task 3 or the last iteration of this task. You are trying to learn the minimum throttle that will allow the plane to fly. If you are flying level, then keep doing the exercise at this throttle level. If you are flying by yourself and launching left-handed, you can use your chin to do this. Figure 5-12 shows this technique.

Figure 5-12    Using the chin to set the throttle.

5. Launch with the intention of flying flat and level for 50 feet at 5 to 10 feet of altitude. Land remembering to cut throttle.
6. Some things to consider:
    a. The Flack will be much more responsive than it was in glider mode. Smaller inputs from the sticks will be sufficient to correct the flight path.
    b. If the plane is trimmed properly and launched cleanly, the only thing the pilot should need to do is cut the throttle and flare the landing with some up elevator. Let the plane fly itself. The controls are there to correct upsets and change course.
    c. If you are having difficulty, then back up to the previous tasks to reinforce basic skills.
7. Put down transmitter and retrieve the Flack.
8. Repeat until controlled, level flight is achieved for 50 feet. Note the throttle sound, feel in your hand the stick setting that achieves this. Do not fly higher than 20 feet. This will keep the Flack out of trees and limit the energy of crashes.

At this point you can fly, but you cannot turn. Hopefully, you have learned skills that minimize the impact of crashes, such as keeping the nose up and cutting throttle before landing/crashing. Now is an excellent time to take a break, recharge the batteries, and get a good night's sleep so that your neurology can have some time to process all the skills it has picked up.

## Task 5: Turning

Once you have mastered straight flight, you are ready to incorporate turns. Airplanes do not turn like cars or boats. Input right aileron and the airplane will roll to the right and the nose will drop because less lift is being generated, so the plane begins to spiral down. The plane is turning, but it is also diving. Add some up elevator to keep the nose level, and the turn is level. Throughout the turn, the pilot is subtly adjusting pitch and roll commands to keep the turn level and progressing. Be sure to read the roll-reversal section below. The sequence of turning advancement I suggest is as follows:

1. Turn 180 degrees and land. Repeat until mastered. Stay below 20 feet of altitude.
2. Turn 360 degrees and land. Repeat until mastered. Stay below 20 feet of altitude.
3. Link 360-degree turns while keeping altitude constant.
4. Fly figure eights at constant altitude.

Once figure eights are possible, then basic control of the airplane has been established. It is up to the pilot where to continue improving.

### Roll Reversal When Flying at the Pilot

When the airplane flies at the pilot, roll control is reversed. Left roll input results in a right roll from the perspective of the pilot on the ground. This causes a lot of problems. The most common symptom is the *death spiral*, where the airplane spirals into the ground because the pilot doesn't recognize the roll reversal. Some techniques for handling this include

1. On recognition of the death spiral, apply opposite roll control. Don't think about left/right, just do the opposite roll input of what you were doing—doing left, do some right; doing right, do some left. If what you just did made the spiral worse, then do the opposite.
2. When the plane is flying at you and a wing drops, think of supporting the low wing with the stick by putting the control stick under the dropping wing. Practice flying at your head in the flight simulator.

Experienced pilots do not think explicitly about left and right when flying at themselves. They have a mental model of what the plane is doing independent of their orientation. They can be reduced to death-spiral newbies if they lose the mental model, which is very easy to do when flying at night or far away.

### Landing

Note that when you are able to turn, you have the choice of how to land the plane. Land into the wind, just like when you took off. It is much easier on the plane and the pilot.

### The Wind

Show up with a kite, and there will be no wind. Show up with a new airplane, and the wind will be howling. You could try showing up with both a kite and a Flack, but it might anger the gods. The general principles of flying in the wind are as follows:

1. If it is calm, then launch and fly in the direction with the fewest obstacles/people.
2. If it is slightly breezy, meaning that your hair gets blown around a bit, then place yourself, if possible, in such a way that you will launch into the wind. But if people/obstacles are an issue, you can get away with a crosswind or a downwind launch. Be prepared to use more throttle and up elevator to compensate.
3. If the wind is brisk, blowing your hat off your head or making coats flap, then launching into the wind is the only option.
4. If you have to lean into the wind to keep standing, then go home.

The wind also can produce lots of turbulence as it goes over trees, buildings, and other surface features, and this means that the airplane will need much more correction from the disturbances. It makes flying harder and more fun. Flight simulators do not do a good job usually of simulating turbulence. Some other points to remember include

1. When carrying airplanes in the wind, it is easiest to "fly" them into the wind using the launch-position grip—they are naturally stable that way. If you fight the wind, the Flack can be damaged.
2. When flying in stiff winds, be very careful to keep the Flack upwind of you at all times; otherwise, you may not be able to get the airplane back to you.
3. Land into the wind, if at all possible. It slows down the ground speed of the airplane and makes it easier to land.

### Overall Good Strategies for Learning to Fly

The two most useful things you can have when learning is an experienced pilot to help you and access to a flight simulator. But it is possible to do without. Some other helpful hints include

1. Keep the plane out in front of you and not too high or too far away. An immediate corollary is that flying over your head or behind you is difficult.
2. Launch with a fight plan, such as "fly three circles and reverse direction."
3. Turn both ways—you will find that you have a preference, but don't allow it to dominate.
4. Keep low and slow in the beginning—less than 20 feet of altitude. This is the opposite advice from traditional teaching—you often hear "keep three mistakes high" when learning. This makes sense with fragile balsa models, but the Flack is designed to take crashes. Being high just increases the energy of those crashes.
5. Be easy on yourself. Flying an RC airplane requires odd skills that are independent of flying airplanes. Every actual pilot that we have taught reported that it was harder to fly RC than a real airplane in certain respects. We have had fighter pilots struggle to control the Flack because left/right orientation is so difficult to process. This is going to take a while.

## Skill Building

Flying skills come from experimentation, risk taking, and establishing consistency. The single biggest thing that will advance your skills is to declare a flight plan and execute it, for example, "one left turn with a loop before doing a right turn three times." Some skill builders that are fun include the following.

### Precision Landing

Confidence that the plane can be placed in a 4-foot square really helps at tight flying sites. Land the plane at your feet, and work your way up to hand catching the plane. This is easier with a good wind and impresses people to no end. If you are flying in a park with some nervous folks who can't tell if you are any good, then a hand catch will establish that you have lots of control. Be careful, missing the hand catch means a face full of airplane.

### Hit the Balloon

A hugely fun drill is to get a Mylar helium balloon and put it with a weight on the end of its string in the middle of the field. Try to hit it with your plane. It is surprising how difficult this can be. It forces the pilot to fly low, maneuver aggressively, and plan ahead. Bring a crash kit.

### Fly in Stressful Situations

There is a world of difference in flying over water as opposed to flying over land. I flew a Flack off a sailboat under sail with a camera mounted on it. The flight was stressful and a failure, but it was a good lesson. Such a flight forces the pilot to be conservative about flying. Flying over rooftops at night is another stressful situation that takes some guts. Fly in conditions that are comfortable but with consequences that are not comfortable. Hurting animate beings is not a good consequence, so do take extra care around persons or animals.

### Aerobatics

Loops and rolls are fun and easy. Flying inverted takes more skill and can be nerve-racking at low altitudes. The Flack is not a precision aerobatic aircraft, but it loops, rolls, and flies inverted just fine. You can learn a lot by working with its capabilities. Some maneuvers worth trying and perfecting include

1. *Loops.* The basic loop is very easy. Just go to full power and give full up elevator, and the plane should loop nicely. If the elevons are not matched, the airplane will roll some during the loop. Balance your up elevon throws to fix this. As you get better at looping, try to make your loops perfect circles and the exit from the loop at the same altitude and heading as the entry.
2. *Rolls.* Rolls also are quite easy in their basic form. Just apply full right or left aileron and watch the plane corkscrew around. Doing it well requires subtle application of the down elevator for the inverted portion of the roll or the nose will drop. Making rolls axial requires considerable skill.
3. *Flying inverted.* Inverted flying can be initiated by a half-roll or half-loop. It is not very easy to maintain, so be sure to give yourself sufficient altitude to recover with a half-loop. Once inverted, the plane will likely require substantial down elevator to maintain level flight. Up and down are reversed, but roll control remains the same as with normal flight.

## Conclusion

Learning to fly takes patience, but anyone can do it with enough practice. Only fly for a half hour a day until turns are solid. Let you brain rest, and sleep on the new skill it is getting. Most of all, keep it fun. Chapter 6 covers how to diagnose and repair your airplane.

# CHAPTER 6

# Keeping You and Your Airplane Alive: Diagnostics and Repair

Planes crash, equipment fails, and dogs eat airplanes. It is all part of living on the bleeding edge of the do-it-yourself (DIY) movement. This chapter takes you through the process of rooting out the problem and fixing it. Remember that your airplane was built from scratch, and there is nothing on it that you cannot fix or replace. You are in a powerful position to sort out problems.

Airplanes tend to degrade before something catastrophic happens, so this chapter covers several quick inspections that help to verify that everything works and is in alignment. It then goes into details about the contents of a crash kit, diagnosing problems, and fixing them. It is the knowledge from this chapter that keeps planes such as that in Figure 6-1 flying. If you get truly stuck, then send me an e-mail (bible@brooklynaerodrome.com).

## Preflight Checklists of Use

### The Brief Preflight in 3 Seconds:
### Servos, Up, Down, Left, Right, Power, Launch

Before every flight, I test the elevator and ailerons for movement before applying power and launching. This mirrors my routine of checking for phone, wallet, and keys before I leave the apartment every day. Steps include

1. Servos move at the same time with circular movement of the stick.
2. Verify up.
3. Verify down.

**Figure 6-1**  A Flack that has seen a lot of flying and is still in rotation.

4. Verify left.
5. Verify right.
6. Run up motor.
7. If all is okay, launch.

The first five steps are an old habit from flying remote-control (RC) gliders. True story: I was hanging out with Joe Wurts, then world champion in RC gliders, and he launched his F3-B Diamond (superfancy, custom world-champion-establishing 2.5-meter design) off Parker Mountain, having forgotten to turn the plane on. A servo check would have prevented the crash—Joe, in his characteristic "keeping it real" approach to flying, did not care if I shared the story.

The quick check is a great habit to have in general. It can reveal stripped servos, broken control horns, and a myriad of other conditions that would lead to a crash if the plane were launched. Look at the elevons as they move. Notice any stutter that may indicate a stripped servo. Perhaps only one elevon moves or only in one direction (broken control horn). Looking hard before launching also will catch servo wires about to be cut by the prop, battery packs about to fall off, etc. On application of power, pay attention to excessive vibration or lack of thrust. Then launch if all is well to see if you were right.

### More Complete Checklist for Aircraft of Questionable Integrity

Below is the checklist I go through before I fly a new plane, a freshly repaired plane, someone else's plane, or a plane that is not flying as I expect. This checklist assumes that servos move and are not stripped and that the motor spins the prop. The inspection starts with a visual/physical inspection with the plane unpowered and then a powered-up inspection. Once you get the hang of it, the inspection takes about 60 seconds and is well worth it.

### Visual/Physical Inspection

With the plane powered down:

1. Check the nose for excessive damage. The Flack can have a pretty torn-up nose, as shown in Figure 6-1, and still fly fine, but a little tape goes a long way.
2. Check that the stabilizers are attached solidly and that the tabs have not torn off. Fix as needed. The stabilizers can be taped on in the field, but it makes them hard to remove for transport.
3. Attach the battery to the airframe, but do not attach the power leads. Make sure that the Velcro on airframe is secure. Fix as needed.
4. Verify the center of gravity (CG) and fix if needed. If the nose is really chewed up and/or the CG mark is illegible, then use ½ inch in front of the prop hole as the CG point instead of 10½ inches back from the nose.
5. Holding the plane in a launch position, make sure that there is no noticeable droop from the nose to aft of the prop hole. A droopy Flack will have very poor up-elevator response because of the forward-pitching moment brought on by all the camber (curve to the wing) evident in the tired airframe. See Chapter 10 for more about this. It is also discussed with regard to the Flying Heart plane in Chapter 9. Fix by stiffening the airframe fore to aft with bamboo skewers or other means, but be ready to replace the wing soon.
6. Verify that the servo arms have retaining screws in them. I have forgotten to add screws many times after a new build or repairs. Luckily, I have never crashed as a result.
7. Gently move the servo arms by hand to feel for any skips that indicate stripped gears. Go to the stripped-servo section below for fixes.
8. Check that the elevon hinges are tight. The tape hinges can work loose and may need to be redone. Use of a higher-stick tape such as duct tape may be called for because packing tape does not stick well to bare foam once it has lost its plastic sheathing. Loose elevons are another indication that the airframe needs replacement.
9. Check that the deck is well attached to the airframe. Fix with zip ties or tape.
10. Check that the motor mount is solidly attached to the deck. Fix with additional zip ties or Coroplast as needed.

11. Check that the angle-stock motor mount is not bent out of shape. Fix by bending back.
12. Check that the motor is solidly mounted to the angle stock.
13. Check that no wires can get drawn into the propeller. Attach loose wires with zip ties. Tape is not a good way to keep wires secure because it tends to lose grip over time.

Next is the powered-up inspection.

### Powered-Up Inspection

1. Power up the transmitter, throttle down, and power up the airplane.
2. Verify control directions:
   a. Verify that the up elevator moves both elevons up and that the down elevator moves both elevons down. If not, reprogram the radio or verify the elevon mixing setup, as discussed in Chapter 4.
   b. Verify that left aileron input moves left aileron up and the right aileron down. Right aileron should do the opposite. If not, and the elevator is correct, then you can fix the problem by switching the servos in channels 1 and 2. Often, after repairs, servos get put into the wrong channel. I always check left and right looking from the back of the airplane. This also allows me to check that the two surfaces center to the same spot after left and right inputs. If not, go to the section on servo/elevon problems.
   c. Verify elevon trim. The place to focus is whether the elevons have the expected ¼-inch up trim or the established flight trim of the airplane. Possible problems and solutions include
      i. The elevons have excessive trim that cannot be made correct by the elevator/aileron trim tab on the transmitter.
      ii. The output horn on the servo may have slipped one or more splines forward or backwards, as indicated by the servo horn not being at 90 degrees to the deck with the elevator and aileron trim tabs at neutral. This happens often with HXT900 servos after a crash or hard landing. See Chapter 4 for how the verticality of the servos is intended. The fix is to remove the retaining screw of the servo output shaft and recenter.
      iii. The deck has slipped its moorings. If so, try to reestablish the correct position with tape and/or zip ties.
3. Verify motor. Run up motor to full power while holding the plane in safe position that allows the prop to turn freely.
   a. If motor does not produce the expected thrust (about 1 pound for our typical motors):

      i. Be sure that the propeller is not on backwards. Raised writing on both the APC and GWS props should be toward the front of the airplane.

    ii. Verify that the battery is fully charged.

   iii. Make sure that the propeller adaptor/prop saver is tight on the prop shaft.

   iv. Go to motor diagnostics section below if still problematic.

a. If the thrust is going backwards, reverse any two of the three motor leads from the speed control to the motor. Power down the airplane before doing this.

b. If the airplane vibrates more than usual, look for

      i. A broken or split prop. Replace.

    ii. A propeller that is not well seated on the prop saver. Readjust and/or replace the O-ring or rubber band. You can use a hair holder or zip tie as a quick field repair.

   iii. A bent prop shaft. You can see this most easily by removing the prop and seeing whether the motor shaft is bent as you rotate it. Fix by replacing the prop shaft or the entire motor.

The last issue with an airplane of questionable integrity is whether the observed wonkiness is due to radio-interference issues. This can be very difficult to diagnose. See the section on poor radio connections to address this.

Next up are supplies to take to the flying field/roof/parking lot/street that you call your aerodrome.

## Crash Kit

A crash kit contains tools and spare parts that enable field repair. Given the simplicity of Brooklyn Aerodrome aircraft, anything can be fixed or replaced at the field. Below is an annotated list of parts and tools for a minimal crash kit. A more complete list follows.

### Basic Crash Kit

1. Two spare propellers—these break all the time; if nothing else, always have a spare prop or two.

2. If you have a prop saver, then you need to have some rubber bands or O-rings—prop savers tend to break or go flying on landings.

3. A spare prop adaptor if you have a screw-on prop adaptor—they tend to unscrew themselves on landing if the power is on, and the adaptor nut is lost in the grass.

4. Needle-nose pliers with cutters—useful for adjusting trim, cutting zip ties, and tightening prop nuts.

5. Small Phillips screwdriver that fits the servo-horn retaining screw—for adjusting the output arm and punching the hole though the Coroplast and airframe.

6. Medium screwdriver that fits prop saver screws.

7. Allen key if the radial motor mount uses an Allen set screw.

8. Box cutter—you need to be able to cut things.

9. Ten zip ties—good for firming up loosening elevon control horns or decks.

10. Roll of packing tape—good for repairing wrinkly noses and busted stabilizers.

11. Spare batteries for the transmitter if it uses nonrechargeable batteries—it's no fun to go to the field only to realize that the transmitter is dead.

12. One spare servo with predrilled servo arms for coat-hanger push rods—servos strip, and arms break.

13. Headlamp—vital if flying at night.

14. Eight-inch Velcro tape—Velcro can come loose on the battery or the plane, particularly if it is cold.

15. Bind clip for receiver if 2.4 GHz—receivers can mysteriously lose their bind to the transmitter.

16. Two paper towels—enough to dry the electronics of a wet airplane.

17. Sheet of paper—it is nice to have a place to jot down notes, share e-mail, etc.

18. A sharpie for taking notes and marking foam.

19. Box for the preceding. Figure 6-2 shows the Brooklyn Aerodrome minimal crash kit. Note the crash-kit contents list taped to the inside of the lid to help remind us to keep it stocked. This sheet is also good to write down items that were needed but not in the kit based on experience.

20. A commitment to only use the tools and parts in the crash kit in the field.

21. Government-issued ID for the nice police officer if he or she requests it.

## More Complete Crash Kit

The contents of a more complete crash kit are listed below in order of likelihood of being needed—every electrical part of the typical Brooklyn Aerodrome airplane is listed. It is very convenient to be able to diagnose problems by replacement rather than trying to debug other ways.

1. First aid kit for human crashes, including
   a. Sunblock
   b. Chapstick

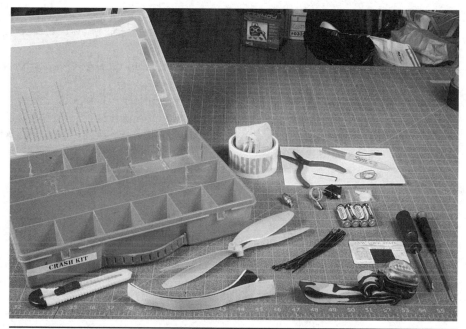

**Figure 6-2**  Basic crash kit and contents.

    c.  Finger-sized bandages for propeller cuts
    d.  Sterile pads
    e.  Painkiller (e.g., aspirin, ibuprofen, etc.)
    f.  Bug repellant
    g.  Antibiotic ointment/spray
    h.  Drinking water
    i.  Emergency snacks

2. More props (six)
3. More servos (two)
4. More zip ties (40)
5. Tree extraction kit (arbor throw weight plus line)—this is a good $20 investment for getting planes out of trees.
6. Skydiving shaded goggles that fit over glasses—for when someone forgets sunglasses
7. Spare motor—motors can burn up and prop shafts get bent.
8. Spare electronic speed control (ESC) with all connections in place—ESCs fail less often than motors unless things get wet.
9. Spare prop adaptor/prop saver—these items love to spin off and get lost in the grass.
10. Bamboo skewers—useful to firm up floppy airframes or decks

11. Duct or gaffer's tape—some repairs need a stickier/heavier tape than packing tape. Examples include nose repair and elevon hinge replacement.
12. Two spare stabilizers—generally, stabilizers will survive with repairs, but it is nice to have a spare in any case.
13. Spare receiver—rarely an issue unless the plane has gotten wet
14. Spare transmitter—I have never had a transmitter failure in the field, but you never know. I always have a spare for paid appearances, Maker Faires, and the like.

## Kinds of Repairs

### The Big-Crash Postflight Inspection

So you stuffed it, and there was a crunch. Most likely everything is fine, but a more careful process is called for. Below is my current checklist that I use in my classes:

1. Do not move the plane until you are sure that you have
   a. The battery pack
   b. The propeller adaptor, O-ring, propeller, or other parts that may have flown off
   c. Not moving the plane is especially important if you are flying at night.
2. Make sure that the battery is not puffing up or smoking.
   a. If you see any problems, do not touch the battery—it will likely be very hot.
   b. Disconnect it, if the battery is cool.
   c. Fifteen minutes is a reasonable "extrawatchful" period for a whacked battery. Do not put whacked batteries in your pocket or other bad places for a fire to happen.
   d. If the battery is burning, there is little that can be done to stop the fire. In my limited experience, water does a reasonable job of reducing the intensity of the flames and does not make things worse.
3. Once you have your airplane and affiliated debris, clear the field as quickly as possible if other people are flying.
4. Power down the airplane and then the transmitter.
5. Do the "More Complete Checklist for Aircraft of Questionable Integrity" above.

## More Specific Fixes

### Completely Dead Airplane

This is a detailed flowchart-style series of steps for diagnosing an airplane that doesn't respond at all. If it is a new build, double-check that the instructions have been closely followed—particularly the polarity of the speed control and servos to the receiver. If the airplane landed inverted in saltwater and marinated for 30 minutes, then most likely all the electrical components are fried, so the incremental approach below will just be frustrating. It is better to order new electronics and focus on seeing if any of the components survived after being handled by the "Water Immersion" section below by substituting them one at a time into a working airplane. Another test is to smell the ESC, motor, and receiver to see if the electronics smell burned. This may help to point the way to a suspect component.

This sequence of steps is meant to apply where the airplane has stopped working after a crash or other mysterious circumstances. It works back from the battery pack to the servos. Unfortunately, it is based largely on incremental replacement of parts that you may not have. Without replacement parts, a multimeter can help to diagnose problems.

The definition of *dead* assumes that

1. The transmitter is on.
2. The throttle is down.
3. The speed control is connected to the receiver on channel 3 with proper orientation as per receiver instructions (see the "Powering Up the Airplane" section in Chapter 3).
4. Servos on channels 1 and 2 have correct orientation as per receiver instructions.
5. The airplane battery is freshly charged.
6. When the battery is connected to the speed control, the servos do not move with pitch/roll commands and the motor does not turn with increased throttle.

The sections that follow lay out the more probable to less likely reasons for a completely dead airplane.

### Completely Dead Airplane: Rebind the Receiver

The number one reason for a completely dead airplane is that the receiver is not bound to the transmitter. The 2.4-gigahertz receivers need to be bound to their transmitter to function. When unbound, they do not listen to the signal from the transmitter and make the airplane look completely dead. Receivers can become unbound for mysterious reasons. The fix is to follow your radio procedure to bind the transmitter to the receiver. Instructions for Hobby King/Fly Sky six-channel radios are included below because they typically ship without a manual. Most binding processes have some type of indicator light. If you don't see the indicator light (it is in step 3 for binding the Hobby King/Fly Sky radio below) and you are confident that the ESC is connected correctly to the receiver, then proceed directly to checking the battery connection of the ESC.

### Binding Instructions for Hobby King Hkt6a/Fly Sky FS-CT6B Class Radios

Starting with the transmitter and the airplane powered down:

1. Remove the propeller.
2. Put the bind clip into the battery channel of the receiver.
3. Power up the receiver. You should see a rapidly blinking red light-emitting diode (LED) from within the case.
4. The binding transmitter should be within 1 yard of the receiver, and no other 2.4-gigahertz radios should be on and nearby.
5. Press the Bind button on the transmitter.
6. Power up the transmitter while still holding the Bind button. Keep holding the Bind button until the blinking LED on the receiver stays on.
7. Power down the airplane.
8. Remove the bind lead (I often forget to do this).
9. Turn off the transmitter.
10. Turn on the transmitter.
11. Power up the airplane.
12. Within a second or three, you should have control of the servos and motor.
13. Continue to the next section if this did not work.

### Rebinding the Receiver Did Not Work

Depending on how this failed, you may have different problems.

1. If there was no blinking light when binding was done, go to the next section, "ESC Not Sending Power to the Receiver."
2. If there was a blinking light, you know that the ESC is powering the receiver. Go to the section on evaluating the receiver.

### Completely Dead Airplane: ESC Not Sending Power to the Receiver

*What is known:* Receiver not getting power from the ESC.

*Goal:* Determine whether

1. The battery is bad.
2. The battery connector on battery side is broken.
3. The ESC connector to the battery is broken.
4. The battery eliminator circuit (BEC) is fried.

This section assumes that no power is getting to the receiver via the diagnostic of the rebinding process. Most receivers have some sort of blinking LED that indicates that it is "listening" for the transmitter. Alternatively, a multimeter reading 0 volts off the positive and negative leads of the receiver plug of the ESC also means no power. Proceed as follows:

1. Follow the steps of the "Check Battery Pack" section below. If battery pack passes, go to next step.
2. *Known:* Battery pack is good. Do the following:
   a. Inspect the ESC battery connector. Visual inspection of the solder joint may require removing the heat shrink from around the ESC connection or other efforts. Resolder the connection if it looks sketchy at all. This is a common failure point. If the solder joint is bad, sometimes flexing the wires will establish the connection intermittently.
   b. If possible, verify that solder joint is good with the continuity tester on a multimeter or measure voltage from the ESC power leads.
   c. Test to see if the airplane works; if not, replace the ESC because of a defective BEC.

### Completely Dead Airplane: Receiver/Servo

I have had several receivers die from being immersed in saltwater, physical damage, and maybe having a battery reverse-connected to the ESC. The only nonspecialist way to test a dead receiver is to replace it with another one and see if that fixes the problem—remember to bind the replacement receiver to the transmitter. Steps include

1. *Known:* ESC is good or assumed to be good.
2. Connect the ESC to the receiver on the throttle channel.
3. Connect one servo to the receiver—either channel 1 or channel 2.
4. Power up the ESC.

5. If there is no servo "jump" with application of power, then get a different servo, and retest for a servo jump. Digital servos typically will not jump, so this test does not work with them.

6. If there is still no servo "jump" with the different servo, then replace the receiver, and start over with step 1.

7. If the new servo "jumps," then replace the old, nonresponsive servos, and start over.

If there is still no "jump" of the servos when the flight battery is attached, revisit the preceding steps because at this point the whole battery-to-servos unit of the airplane has been replaced.

## Completely Dead Airplane: Transmitter

If after replacing the ESC and receiver the plane is still dead, then the transmitter is suspect. It is time to check out the manual and/or get more specialized help. If at all possible, try a different instance of the same brand/model transmitter to isolate the problem. Remember to rebind the receiver if the transmitter is replaced.

## Water Immersion

Many a Brooklyn Aerodrome airplane has found itself in water. There is a mud puddle in the middle of McCarren Park after every rainstorm that has an powerful attractive force for delta wings. The following are the steps to take on a water landing:

1. Get the plane out of the water as fast as possible.

2. Disconnect the flight battery.

3. Try to determine how much water may have gone into the speed control, receiver, motor, and servos.

4. Mop up any water on the electronics with paper towels from the crash kit—the goal is to keep the water from spreading and making things worse.

5. Any component that has gotten saltwater in it should be removed from the deck and plunked in freshwater immediately. The salt in the saltwater will short out the components. Freshwater generally will not hurt electronics. If the components got a good dousing in saltwater, they probably destroyed themselves immediately, but not always. Store the rinsed components in a warm place for 24 hours to dry, and then test them.

6. If the water was fairly clean, the plane may only need to sit in the sun for a few hours to dry out. I have successfully flown completely wet airplanes, but doing so risks destroying components—ESCs seem particularly vulnerable. I dunked an entire plane in the lake in Prospect Park in Brooklyn after a motor flameout and had all the electronics survive, including the high-definition (HD) camera.

## Servos Operate But Motor Does Not

This is a fairly common problem that has several possible causes:

1. The cutoff voltage has been reached for the ESC, and the motor power has been cut to protect the batteries from overdischarge. Verify by trying a different freshly charged battery pack. If that solves the problem, then nothing is broken—go charge your batteries and fly.
2. Did you have zero throttle and the throttle trim set all the way down on the transmitter (which was on) when you powered up the airplane? Most ESCs lock out the throttle if the plane is powered up with nonzero throttle. Speed control might be chirping like crazy because of the dangerous situation (the motor otherwise would start turning). Test by powering up the airplane with zero throttle from the transmitter.
3. Some ESCs require that the throttle be armed. For example, Jeti ESCs require the throttle to be at zero, then go to full throttle until a series of beeps is heard, and then go back to zero to arm the motor. Read your manual—cheaper speed controls rarely have this feature.
4. Verify that your ESC is on the correct channel (usually channel 3) for how the radio is set up.
5. Verify that the throttle is sending a signal to the throttle channel by putting a servo on that channel and the ESC to any other available channel. If the "throttle" servo moves in accordance with the throttle stick, then you know that the radio is working properly with the receiver. If not, then investigate channel assignments and/or programming in your radio.

If the problem persists, then either the motor or the ESC is broken. If the flight battery has been connected backwards to the ESC, then most likely the circuits have been destroyed, and the ESC has to be replaced. Sometimes the ESC survives, however.

1. Inspect the motor for damage, as described in the motor section, and replace if necessary.
2. If the motor still does not turn, then swap out the ESC with a replacement.

## Evaluating Motor Damage

The outrunner motors we use at Brooklyn Aerodrome are remarkably robust and powerful, but they can break, overheat, and degrade. This section covers diagnosis and fixes for them.

### Symptom: Motor Has Weak Thrust

The recommended motor-propeller combinations should generate at least 1 pound of thrust from a fresh battery. You can test this by having a helper put the nose on a scale pointing downward and subtract the weight at no throttle from the weight at full throttle to get an idea of the thrust. Common causes of weak thrust include the following:

1. Is the prop facing in the correct direction? The raised lettering on APC and GWS props should face forward.
2. Is the prop saver firmly on the motor shaft? Check that the prop and motor are solidly attached. Tighten screws if necessary, being careful not to strip the aluminum threads. If the prop saver comes loose often, then grind a flat spot onto the shaft to help keep the screws from slipping.
3. Is the battery fully charged? It is easy to forget to charge a battery. The battery should have 8.4 volts for a two-cell pack.
4. Does the motor move freely unpowered? Hand turn the motor to see if the bell housing is rubbing on anything or if the backside of the output shaft is rubbing on the motor mount. Bearings can get sticky or wear out after many hours of flying.
5. Does the motor show signs of overheating? Resolve the overheating issues by going to the "Motor Gets Very Hot" section below. If you can still fly, get a smaller prop or use less throttle when flying.
6. If the motor runs very roughly or haltingly, then one of the power leads may be disconnected. Check the wires and bullet connectors for bad solder joints, loose fit, or shorts.
7. If problems persist, then try swapping the motor out. If that makes no difference, then the ESC is the next candidate for being the cause of the problem. See the section on evaluation and repair of the ESC.

### Motor Vibrates

These motors can exceed 10,000 revolutions per minute (rpm), which translates into a lot of vibration if the moving parts are not balanced. Following are the common sources of excessive vibration:

1. Check to see if the propeller is broken or split. Go to the "Broken Propeller" section below for fixes.
2. Verify that the motor is solidly connected to the motor mount. The screws and zip ties holding the motor on can get loose or break. Fix by tightening up all loose components.
3. See if the propeller shaft is bent. This can be quite subtle. Things to try include
   a. Visually inspect the shaft while rotating the motor by hand to see if there is a noticeable wobble.
   b. Remove the prop, and run the throttle up so that the motor is turning at a low speed. Carefully put your finger on the shaft and see if it is vibrating. This indicates a bent shaft.
4. If the shaft is bent, you can try to bend it back, which can be quite difficult. Replacement shafts are available for motors, but I tend to just replace the whole motor if bending the shaft back doesn't work.

## Motor Gets Very Hot

Overheating will degrade the insulation of the windings, and they will discolor. If there are signs of damage, it is time to order a new motor. Heat also damages the magnets. If after a flight you cannot leave your finger on the bell of the motor for 5 seconds, then the motor is being run too hot. Options for overheating motors include

1. Get a smaller prop, or go down in pitch. If you are flying a 10×4.7, consider dropping to a 9×4.7 or a 10×3.8.
2. Use less throttle as you fly. I regularly overprop my motors with the knowledge that they will tolerate brief bursts of full power for launching and emergencies but mostly fly at one-half to three-quarters throttle.
3. Get a bigger motor. While it can be a hard to figure out what a bigger motor is across brands, within brands the relative power of the motors is generally evident. Be aware that you may need to upgrade the speed control to handle more amperes as well.
4. All the designs at Brooklyn Aerodrome have excellent airflow to the motor. But if it is enclosed, the motor can get hot as a result of insufficient airflow.

## Motor Doesn't Work at All

A nonresponsive motor can be due to any number of things.

*What is known:* Servos are working normally; motor is not moving at all.

*Goal:* Check for physical damage to the motor.

1. Is the insulation on the windings blackened or discolored? If so, the motor has overheated to the point that the insulation on the windings has melted, and the motor is shorted out. Replace the motor. It is likely that the ESC was destroyed as well, so be prepared to replace that also.
2. Did the motor suddenly stop working with a puff of smoke coming out? Then it shorted out and needs replacing, as is likely for the speed control.
3. Check the bullet connectors on both the motor and the ESC for being loose or broken.
4. Check the motor wires for damage. If damaged, repair them.
5. Proceed to the next step if no damage is found.

*Assumption:* Servos are moving with transmitter inputs, but motor is not turning at all. Motor has no obvious physical damage.

*Goal:* Verify that the throttle control is driving the ESC.

1. Power down the airplane. Remove the propeller.
2. Move the ESC from the throttle channel to any other channel.
3. Move one servo from its normal channel to the throttle channel.
4. Put the throttle down, power up the transmitter, and power up the airplane, being ready to disconnect the power to the airplane if the motor starts spinning. If motor spins, then the throttle channel is not configured correctly. Consult your radio manual.
5. Move the throttle, and note whether the servo in the throttle channel moves as well. If it moves, then you know that the transmitter is programmed correctly. If it does not move, then consult your radio manual for how to use the throttle.
6. You have verified that the throttle channel is working. Restore the receiver connections, and go to next step.

*What is known:* The throttle channel is working, the servos are working, but the motor is not turning.

*Goal:* Determine whether the motor or speed controller is damaged.

---

**IMPORTANT** *Never connect a known shorted or electrically damaged motor to an ESC (speed controller) because the shorts can destroy the speed controller. A damaged speed controller is very unlikely to damage a motor.*

1. Replace the nonturning motor with a new one. If the new motor turns, then the problem is solved. Remember to disable the nonturning motor

because the motor is damaged. One way to not use the motor by mistake is to wrap the bullet connectors in tape. If the new motor does not turn, then do not disable the old motor, and proceed to the next step.

2. If the new motor still does not turn, then replace the ESC with a new one. The motor should turn now because the ESC and the motor have been replaced. Go to next step to determine whether the motor should be discarded.

*What is known:* The ESC was defective and has been replaced. The motor may or may not be defective.

*Goal:* Determine how likely it is that the motor has an electrical problem that could destroy a working speed control.

1. If the motor failed in flight and shows discoloration on the windings or signs of overheating, then discard the motor because the evidence is very strong that the motor is shorted.
2. If the speed control is obviously the source of the problem because battery power was applied with incorrect polarity, then the motor is likely okay.
3. Label the motor leads *A*, *B*, and *C*. With a multimeter, measure the resistance of *A–B*, *A–C*, and *B–C*. If they are not equal, there is a short, and the motor should be discarded.
4. Mount the motor shaft in a drill. Run the drill up to maximum revolutions per minute (1,000 rpm), and measure the alternating-current (ac) voltage of the pairs *A–B*, *A–C*, and *B–C*. If they are not the same, then there is a short, and the motor should be discarded.
5. Test for continuity between each of *A*, *B*, and *C* and the stator. You may have to expose some bare metal with a file or sharp blade. If there is a continuity, then discard the motor.
6. If the motor passes these tests, then there is a reasonable chance that it is not damaged and can be used. Be forewarned, though, that shorted motors likely will destroy a new speed control.

## Battery Damage

Batteries have improved tremendously over the years, and they are far less likely to burn your house down than the early lithium-polymer (LiPo) batteries were. But battery damage is a very touchy subject. Let's break it down into the major damage issues and quick responses:

1. *On-fire battery syndrome.* Figure 1-11 shows a burned LiPo from physical damage. The reaction is very hard to stop, but the best way to deal with it is to let the fire run its course. The flames can shoot a few feet high, so keep your distance, and the fire should be over in 3 to 4 minutes for the size batteries used in these airplanes. Application of water does a reasonable job of reducing the fire's intensity. Sand is a good choice too. See the "Battery Disposal" section below.

2. *Crinkly battery syndrome.* This is the result of high-velocity interactions with Mother Earth resulting in a visibly distorted battery case. If you have crinkly batteries, then please consult the "Highly Suspect Battery" section below.

3. *Puffy battery syndrome.* So the battery looks fat and happy but likely doesn't deliver the juice it used to. All LiPo batteries degrade as a function of time and treatment, so it may not be your fault. What is happening is that the battery is gassing off inside the protective wrapper, which is containing those gases. If the batteries look really puffy (more than 25 percent bigger), then go to the "Highly Suspect Battery" section. Otherwise, keep an eye on the pack and continue to use it normally.

## Highly Suspect Battery

So the battery is either wrinkly or really puffy—what are the next steps? The prudent thing to do is to replace the battery. It is damaged, and very little more damage may make it catch fire. It is also unlikely to be performing very well. Nearly everyone will have a reason to push this limit (batteries are on order, only one charged battery left, etc.). So given that the limits of safety are going to be pushed, let's do it as carefully as possible. Steps for handling suspect batteries include the following:

1. Never leave a suspect battery unattended on a surface that can burn or where something above it can burn. Ideally, the only time it is on a surface that can burn is on an airplane.

2. An immediate corollary of this is that one should always store and transport a suspect battery in a fireproof container. A ceramic flower pot with a lid is a great choice, as are purpose-built LiPo pouches.

3. After flying, always touch test your battery to see if it is really hot. If you cannot keep your finger on the battery for 10 seconds, then it has too much internal resistance and needs to be disposed of. This assumes that the battery used to run cool. If it is a new battery, a hot battery pack is evidence that the battery is being overdischarged.

### Check Battery Pack

If the battery successfully charged then the cells are in reasonable shape. All modern chargers check cell voltage and refuse to charge if the cells are not functioning properly. If the battery will not power the airplane and other batteries will, then the culprit is likely the power connection from the battery. I have never had a factory-soldered battery connector fail but I have had crash-damaged batteries sever the power lead to battery tab connection. If that is the case then dispose of the battery. Checking battery voltage with a multimeter is a good test here as well.

### Battery Disposal

LiPo batteries do not have toxic components and generally are as disposable as any other household garbage, but check your local disposal rules. Batteries should be disposed of with as little energy in them as possible. Suggested steps include

1. Run the battery down as far as possible. The easiest way to do this is to connect a 9-volt lightbulb to the battery and let it convert the energy of the battery into light until the battery has 0 volts.
2. You get bonus points if you check with a multimeter. Getting the voltage to 0 volts per cell is the goal.
3. Toss the batteries out with the regular trash, but check your local regulations first.
4. If you ignore the preceding, then at least don't put charged batteries into the trash. Run them down as much as possible.

## Servo Damage

There are a few sources of failure that fortunately are prettily easily diagnosed. As earlier, we work from the most common kinds to the less common kinds of damage.

### Servo Not Moving

Take the following steps:

1. Verify that the servo is on the correct channel (usually channel 1 or channel 2).
2. Verify that the polarity of the connector is correct.

3. Inspect the servo wire to see if it is cut. Resolder it if it is cut.

4. Try the other servo in the problem servo's channel. If the other servo works, replace the old servo with a new one.

5. If both servos are not working but the motor works, then check the servos for being stripped, as described below.

6. If the other servo is still not working, then double-check steps 1 and 2.

### Stripped Servo Test

After a hard crash, servos can strip their internal gears. Test them as follows:

1. With the airplane powered down, gently move the servo arms with your fingers to see if there are any skips in the gear train. You will feel a little jump or loose spot. Replace the servo or replace the gear set if the servo is expensive.

2. With the airplane powered up, test that the servo can resist force applied in both directions at neutral position. Use about 1 pound of force, and be gentle. If the servo gives up very easily, then replace it. A gear change may not be sufficient because the potentiometer may be damaged.

I have flown many airplanes with mildly stripped servos just because there was no replacement. As long as a crash would not be catastrophic (damage to property/person/animal), then use your judgment.

### Servo Arms Are No Longer Vertical When Trim Tabs Are Centered

This tends to be noticed when the airplane cannot be put back into proper trim with the travel of the trim tabs. The pilot is having to constantly fly with up or down elevator when the plane didn't always fly this way. What may have happened is that on a hard landing the mass of the control rods and elevons may have forced the arms one or more splines farther than vertical. The solution is to remove the retaining screws from the arms and set the arms back to vertical. This is quite common with HXT900 servos. It took me 5 years to realize that this was happening.

### Busted Servo Arm

It happens that a servo arm breaks, especially in the cold. The only fix is to replace the arm. Have a few spare arms drilled for the coat hanger pushrod diameters in your crash kit.

## Broken Propeller

The prop takes a lot of abuse because on every landing it is in danger of being broken. That said, I flew two days at Maker Faire Bay Area landing on asphalt, and I never broke my prop. It ended up being pretty scuffed up, but broken it was not. Props generally break because they are under power when they hit the ground. A prop life-extending skill, then, is to learn to cut the throttle at sufficient altitude that the propeller has stopped spinning or is just spinning from air flow when it hits the ground.

### Kinds of Prop Damage

Not all prop damage merits replacement, and some repair is possible. Go through the following steps:

1. Prop broken at the base, only one blade. Replace prop.
2. Prop missing a chunk less than ½ inch. You can try replacing the missing chunk with packing tape. If the motor doesn't vibrate badly, then see if the plane can fly. If there is too much vibration, replace the prop. If vibration continues with the new prop, look for a bent prop shaft.
3. Prop split at end. This is generally fixable with a bit of packing tape.
4. Prop bent back but otherwise fine. Gently bend the prop back, and test at full throttle with protective goggles on. The blade may fly off, so take care. If it holds together, then fly.
5. Scuffs, dings, etc. Ignore these unless the motor is shaking a great deal. Replace if needed.

## Flutter

Most model airplanes are much stiffer than they need to be, and as a result, they rarely have issues with flutter at slower speeds. Our aircraft at Brooklyn Aerodrome are less stiff, which gives them great bounce on crashes but can lead to wobbly wings at higher speeds, usually during a power dive. Flutter shows up with either the elevons rapidly going up and down or the wing tips starting to flap up and down. The immediate solution is to reduce power and load the airframe with some G-forces with up or down input. If the airplane has gotten soft from many crashes, then it may be time to consider a fresh airframe.

## Poor Radio Connection

Modern radios should have at least 1,000 feet of range, but various factors can influence the quality of the connection between the transmitter and the receiver. This is more of an issue handled by your radio manual, but some common sources of problems and approaches are covered below.

1. Does your transmitter pass its range test? Consult your manual for how to do this.
2. How much power is the BEC on the speed control expected to provide? If a bunch of night-flying gear is drawing power directly from the receiver, you may be causing low-voltage issues, which, in turn, are causing the receiver to reset. See the "Power Budget" section of Chapter 8 for a discussion of this.
3. Your 2.4-gigahertz radio is competing with a broad range of devices on that slice of spectrum. This may be an issue in urban environments.

## Conclusion

This chapter lays out the problems we have encountered and the solutions we have devised based on our experience at the Brooklyn Aerodrome. The easiest way to fix planes quickly is to have replacement components to try. Components are cheap enough now that having a complete set of replacements is affordable. In addition, they become the gear for the next plane. Be aware that damaged motors can destroy new speed controllers, but that is the only real gotcha in the repair space. Chapter 7 addresses getting your planes looking sharp.

# CHAPTER 7

# Make Your Plane Look Good for Day Flying

O
ur approach at the Brooklyn Aerodrome to building with flat board makes it easy to decorate as well. This chapter lays out all we know about decorating for day flying, and most of that was figured out by Karen, who drives the artistic side of the Brooklyn Aerodrome. Do not decorate your first airframe because it will likely be destroyed in the learning process. If there are kids involved, then some simple decorating on the first plane is a great way to keep it fun while saving the fancy stuff for the second airframe.

This chapter is organized around example airplanes that use various techniques and materials, with a bit of a historic tour of things we have tried. Not all techniques worked out, so read the pluses and minuses of what we have done. Figure 7-1 shows some variations.

## General Considerations to Keep in Mind

Decorating can have both positive and negative effects on the performance of an airplane. The plane can be made tougher, more fragile, faster, or more sluggish by choice of technique and material. Some principles to keep in mind include the following:

1. Decorations add weight. A bare-bones Flack weighs 15 ounces. It can fly comfortably carrying another 8 ounces depending on prop/motor and flying conditions. A better decoration budget would be 5 ounces.
2. The center of gravity (CG) has to be maintained, but 75 percent of the airplane's surface area is aft of the CG, so decoration will tend to move the CG back. The correct CG can be achieved by moving the battery forward, moving the motor forward, or adding nose weight.

**Figure 7-1** Assorted decorated Flacks.

3. Decorate with an eye to how the airplane will look in the sky, not on the ground. Things to consider include
   a. The underside of the airplane will be in shadow most of the time.
   b. Details will not be seen.
   c. Contrast is your friend.
4. Don't worry about clearly indicating left versus right or top versus bottom with color for day flying as long as you are not straying from core Flack design. Night flyers (Crystal Towel) can cause problems, but that is covered in Chapter 8.
5. A flying plane is always prettier than a plane on the bench. Don't obsess on making it perfect.

## Examples

We have tried a lot of decoration strategies at the Brooklyn Aerodrome. Learn from our successes and failures. With a somewhat timeline-sensitive presentation, this is what we have done for the daytime.

### Mouse: Avery Stick-on Labels and Sharpies

Figure 7-2 is the first example of a "decorated" day flyer and shows the genesis of a long series of "face in the sky" approaches that look great in the air. The airplane never flew, but it was the parent of many that did. Our simple technique was sheets of white stick-on Avery label paper on blue foam and various Magic Marker colors.

- *Technique:* Avery stick-on labels and Magic Markers.
- *Pluses:* Maximum control of shapes/lines.
- *Minuses:* Time-consuming; subject to wear and moisture; colors not saturated.

### Purple Monster: Wrapping Paper

The Purple Monster in Figures 7-3 and 7-4 represented a breakthrough not only in design but also in performance. Not only did it look better, but it also flew better in that the wing was made considerably stiffer by the wrapping paper that we used to decorate the airplane. The wing was decorated by the following method, which we still use on all of our sheet-decorated planes.

### Covering an Airplane with Wrapping Paper

1. Cut the deck from Coroplast—later planes stopped using a deck.
2. Cut the wing and stabilizers from blue foam.

**FIGURE 7-2**  Avery stick-on paper and Magic Marker colors for the Mouse.

**Figure 7-3**  Bottom of the Purple Monster.

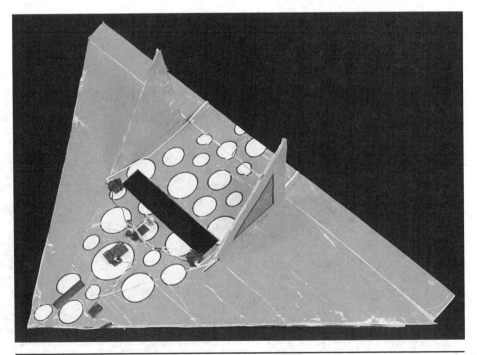

**Figure 7-4**  Top of the Purple Monster.

3. Cut and hinge the elevons with packing tape. It is important to do this before applying any covering.
4. Cover a large, well-ventilated spray area with newspapers.
5. Cut gift-wrap paper slightly oversize for attachment to the bottom of the wing.
6. Spray 3M 77 or other spray-mount adhesive onto the gift wrap. If other adhesive is used, test it on a bit of foam to be sure that it doesn't melt the foam.
7. Wait for the spray mount to get a little tacky, and press the wing onto the gift wrap.
8. Trim the paper flush with the edge of the blue foam.
9. Put down fresh newspaper that is not sticky.
10. Generally, there are two large triangles of material left for the top surface. Apply spray-mount to each. Overlap the center by 1 inch, and apply the paper to the topside of the wing. Trim it flush with the edge of the foam. No effort is made to wrap the edge of the foam.
11. Cut the gap for the elevon. Unless the paper is very lightweight, the stiffness of the paper will resist going into the hinge and interfering with the functioning of the elevon. Trim the paper to the limits of the bevel. The blue foam that is showing can be colored with a Magic Marker.

You now have a wrapping paper–covered wing ready for detailed decoration. Notice that the wing is considerably stiffer than one made of foam alone. Not all wrapping papers will do this. For example, neither tissue, very thin plastic, or vinyl adds much stiffness.

The stabilizers and deck are covered in a similar fashion, but the deck need only be covered on one side. Detailed decorations are added with bits of other colored paper or plastic that are attached with spray-mount. If paper gift wrap is used, then it is well worth considering application of a layer of clear packing tape to waterproof the design. Dew on the grass can ruin a paper-decorated plane. Once decorated, the plane is built as described in Chapters 3 and 4. Note that it is much easier to decorate a plane without the electronics attached. Finish the design, and then add the gear.

### Dealing with Blue Foam Cut Edges

The edges of the blue foam revealed on the sides of the cut and the elevon is often a source of visual incoherence. We have approached it in many ways.

1. Just ignore it.
2. Wrap the edge in black or other-colored tape before applying the top/bottom surfaces with overlap to the adjoining surfaces to lock the tape in. This can require slitting the tape to go around the corners, etc.

3. Wrap the first surface around about ½ inch before applying the second surface. This takes a long time to do and, like tape, requires slitting the wrap.
4. Color the edge with a Magic Marker.

### Dealing with the Elevon Hinge Bevel in Blue Foam

The elevon hinge bevel can be annoyingly prominent in the design. It would be nice to hide that blue streak, but it is surprisingly difficult to do. Things we have tried and what happened include

1. Before applying the topside layer of material, apply a strip of decorative material on the hinge with it doubled back. This can work with very thin material, but it tends to make the hinge really stiff, and the material will pop out anyway.
2. Cut the top material in the middle of the hinge, and push it down into the hinge, making the blue less noticeable. Unless the material is very flimsy, it will become unstuck and, if stiff enough, interfere with up-elevon movement.
3. Paint or color the foam. This can work fine.
4. Just ignore it, and keep the bevel area free of material. This is what we do most of the time.

### Assessment of the Purple Monster

- *Weight:* The decorations added 7 ounces to a 15-ounce airplane for a total weight of 1 pound, 6 ounces.
- *Technique:* Spray-mount-attached colored wrapping paper, black tape, and clear tape.
- *Pluses:* The design is visually strong; wrapping paper made the airframe stiffer, which meant a crisper performance; it was fairly quick to decorate; and the extra weight did not affect performance negatively, although the airplane did fly faster.
- *Minuses:* Wrapping with clear tape is a hassle to protect against moisture; the airplane was not as noticeable in the sky as we would have liked because of color choices; the spray-mount was smelly and messy; and most of the visually interesting bits are in shadow most of the time on the bottom.

### Big Pink Angst: Fresnel-Lens Plastic

This plane represents the angst of creating new designs. As seen in Figure 7-5, it shows more tooth and claw with a shiny holographic skin that popped out

**FIGURE 7-5**  Top view of the Big Pink Angst. Note the absence of a deck. Bottom is similar to the Mouse and the Purple Monster.

much more than the wrapping-paper skin of the Purple Monster. It is probably our most flown airplane in the Brooklyn Aerodrome fleet, having been multiple years at Burning Man, Maker Faires, and summer camps.

### No Deck for the Big Pink

The prime innovation of the Big Pink is that it is deck-free. The covering stiffened the airframe sufficiently to allow the components to be attached directly to the wing once it was covered and completely decorated. Servos are attached with both double-stick tape and zip ties directly onto the skin of the Big Pink.

The decision to eliminate the deck was motivated by seeking weight savings (2.5 ounces) and the benefit of simplifying the build/decorating process. The function of the deck is to allow for rapid airframe replacement, and it provides essential stiffening of the raw blue foam. I did not expect to crash this airplane as a beginner would, and I did not need the stiffening effect because of the adhesive film's properties.

### Pink Holographic Adhesive Film (and Other Colors)

This material is one my favorite skins. Go to the Brooklyn Aerodrome website to see the range of color this material offers. It weighs about ½ ounce a square foot and stiffens foam airframes like no other material we have. Given that the sides of a Flack are about 8 square feet and the stabilizers add nearly an additional 2 square feet, the covering adds 5 ounces to the total weight of the airframe. The plane can be covered in about 1½ yards of the 2-foot-wide material. The Flack is 23 inches from nose to elevon and 41 inches wide, so budget 45 inches to cover the top and bottom of the wing, with the remaining 9 inches for the stabilizers. We apply the material in the same fashion as done with the Purple Monster, except without all the smelly/sticky spray-mount adhesive. One frustrating bit is that bubbles tend to get caught under the plastic. This can be fixed by pricking the bubble with a pin and smoothing the plastic with a credit card. We tend to wrap external edges because it looks much better, but this takes a long time.

The plane handles much more crisply than a regular Flack, but it also flies faster and is more fragile on hard landings or crashes. Whereas the foam by itself may bend or crush, with the plastic skin, it tends to tear. It is also going to be much less forgiving if it hits a person or property. Figure 7-6 shows a yellow and pink version with rounded edges that may be less dangerous.

### Assessment of Big Pink Angst

- *Weight:* The decorations added 5 ounces to a 15-ounce airplane for a total weight of 1 pound, 4 ounces.
- *Technique:* Stiff reflective adhesive film; various tapes; no deck used.

**Figure 7-6**   The Bumble Bee, a Fresnel lens–covered rounded airplane.

- *Pluses:* Increased stiffness results in the best-flying plane in our aerodrome; visually strong.
- *Minuses:* Crashes tend to tear the film and foam; colors can fade in intense sunlight; and increased stiffness makes the plane less safe for creatures or property.

### Silver Shark: Plastic Film 2

The Silver Shark continued the big, toothy mouth theme but with a very different skin (Figure 7-7). The goal was to make an airplane that really caught your attention, and we thought that a reflective surface would be best if we could catch the sunlight in a mirror-like way and flash the plane while doing loops and rolls. Also notice that it has only a single fin—it flies fine with half the usual stabilizer complement. It debuted at the Figment Festival in 2010 on Governor's Island in New York City to great effect.

### Details

The skin is an adhesive plastic that has a little stretch to it and has reflective flecks embedded into it. It cost around $10 a yard at 24-inch width, and it took

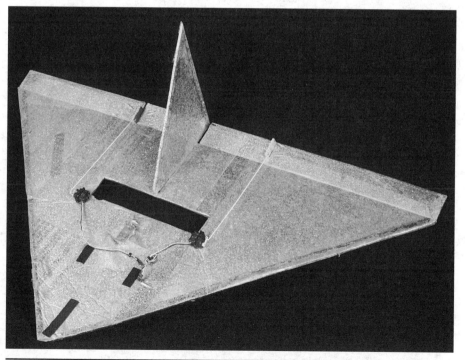

**Figure 7-7**  The Silver Shark with its single terrifying fin. The other side is similar to other mouth planes.

1.5 yards to cover the entire airframe. The material weighs about 1 ounce per square foot. The total covering weight is 9 ounces since the the Flack has 8 square feet of wing and 1 square foot for a single stabilizer. Note that all external blue foam edges were wrapped bottom to top with a ½-inch overlap because we didn't want to lose any opportunity to create a light flash, and the leading edges are an excellent place for that to happen. The adhesive also was a little goopy, so we also wrapped the external edges with clear packing tape for extrastrong bonding.

One downside of this was that with the stabilizer in the center to represent a single shark fin, there was a substantial stress riser in the middle of the airframe that was evident after 30 or so flights. The result was that the Shark got a little soft in the middle and could be seen flexing in loops. The situation could be rectified with a spar the width of the prop hole about 1 inch wide made of Coroplast added behind the prop hole.

### Assessment of the Silver Shark

- *Weight:* 1 pound, 4 ounces.
- *Technique:* Plastic adhesive sparkly covering; various tapes; no deck used.
- *Pluses:* Nice sparkly airplane that got a lot of attention; it flew well, but not quite as well as the Purple Monster or the Big Pink Angst.
- *Minuses:* Center-mounted fin and absence of a deck caused the plane to be soft in the center, which was noticeable in turns and loops.

### Brooklyn Aerodrome Orange Plane (BAOP) Paint on Foam

The BAOP was created to fly at the opening party for the Iridescent Science Center in the Bronx. We wanted a good-looking plane with Brooklyn Aerodrome written on both sides. Figure 7-8 shows how it came out. The colors are safety orange with white lettering. The plane flew great but tended to shed chunks of wing on landing.

### Painting Foam

The foam board we prefer calls out for painting. It is flat, takes paint well, and painting gives it artistic flexibility like nothing else. Unfortunately, the paint removes all flexibility from the surface, which flies great, but is a complete disaster on landing because the airframe breaks into bits on contact with any firm surface. Formulations we have tried include

1. Gesso foundation with colored artists acrylic. *Result:* The airplane broke like porcelain on a stone floor on first landing. Gesso is made of plaster, so no surprise there really.

**Figure 7-8**  BAOP covered in orange paint with white lettering.

2. Artist acrylic applied directly to blue foam. *Result:* Slightly less explosively fragile but still too fragile for robust use.
3. Foam-friendly spray paint. *Result:* Applied sparingly, it appeared to work better than any other paint-based alternatives.

Another issue with paint is that it tends to not allow packing tape to adhere well, making field repairs very difficult. Figure 7-9 shows zip ties being used to stitch on a chunk of wing and nose that would not take tape.

### Assessment of the BAOP

- *Weight:* 18 ounces.
- *Technique:* Deck with orange acrylic artist's paint and stenciled letters.
- *Pluses:* Great-looking plane that flew as well as plastic film–covered planes.
- *Minuses:* Landings had to be perfect or the plane would break into chunks; not a recommended approach.

**Figure 7-9** Crash damage had to be repaired with zip ties because the paint would not take packing tape.

### Blue Angel and Others in Printed Tyvek

We have a long-standing relationship with a fine-arts printer (Skink Ink) that can produce custom skins. The airframes in Figure 7-10 are all printed first and then attached. It is an excellent way to get high-quality images onto foam.

Perhaps the finest example of this method of decoration was a presentation plane created for the Secretary of the Navy shown in Figure 7-10. The level of detail possible with this approach is limited only by the designer's imagination. Figure 7-11 shows the cockpit details.

The material is Tyvek, which is a very tough plastic skin that is ideal for covering foam. Unfortunately, we have not been able to find Tyvek with an adhesive backing, so the covering method is either manually applied adhesive film or spray-mount adhesive, as done with the Purple Monster. Printing costs are around $150 per airframe (it comes already attached to foam) for standard designs. But the results speak for themselves. Custom printing costs more. The printing can be glossy or flat. For presentation planes, this approach cannot be beat.

### Assessment of Printed Tyvek

- *Weight:* 22 ounces.
- *Technique:* Deckless build with contact paper attached custom-printed Tyvek.

**Figure 7-10** Printed Tyvek-covered airframes. (*Left to right*) Custom corporate logo print, Navy Blue Angel–styled plane, and camouflage pattern in orange. (*Bottom*) A dragon design.

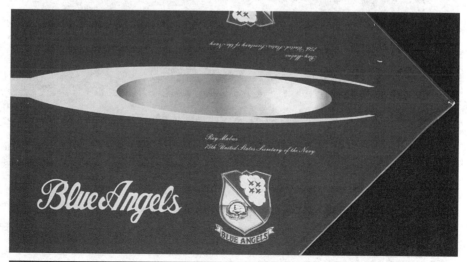

**Figure 7-11** Cockpit details of the presentation plane for the Secretary of the Navy.

- *Pluses:* Complete design freedom that results in a plane that looks as good up close as it does far away; ideal for presentation planes; the Tyvek is a very strong and appropriate covering that is nearly as stiff as adhesive plastic films.
- *Minuses:* Custom printing is expensive.

## Flying Heart: Sign Vinyl

Figure 7-12 shows the Flying Heart in red and white stripes. It is the only example we have of a sign vinyl–covered aircraft, which is why it is included here despite its departure from the classic Flack shape. The details of its construction are covered in Chapter 8. Perhaps the greatest advantage of sign vinyl is that it can be had for free as a by-product of sign making. When you are getting your recycled Coroplast from the sign shop, ask the sign makers if they have any scrap vinyl, and you may find yourself with a garbage bag full of varied scraps. The material comes with adhesive and a removable backing and is very easy to work with.

Bought new, it comes in rolls that can be cut into varying-width tapes, which is what we did for the Flying Heart. It cuts easily with scissors, which makes it an ideal material for working with kids. The adhesive is less aggressive than that of the other plastic films, so pieces can be repositioned without tearing up the foam. Another benefit is that it resists scrapes and does not pick up grass

**Figure 7-12**   Flying Heart covered in sign vinyl.

stains. It is an ideal material for a tough and long-lasting aircraft. The only downside to sign vinyl is that does not stiffen the airframe very much, which means that a deck will be required. It weighs 0.6 ounce per square foot.

### Assessment of Sign Vinyl

- *Weight:* 22 ounces.
- *Technique:* Decked-based build with sign vinyl covering all surfaces.
- *Pluses:* Excellent color saturation, reasonable prices, and easy to work with; extremely tough; takes damage well and is easy to repair.
- *Minuses:* Does not stiffen airframe sufficiently for deckless builds.

### Firefly: Balloon Film

Figure 7-13 actually shows a night flyer with a reflective skin. It was designed to fly indoors at a warehouse party that had a camping theme, so there would be lots of light to reflect in addition to its own electroluminescent wire (el-wire) illumination. From a daytime decorating point of view, the interesting part is the reflective silver plastic, which is a nylon film used to create balloons. It is very light at 0.10 ounce per square foot and can be gotten from http://balloonkits.com. We attach it with spray adhesive, and it is very difficult to get a smooth surface. Over time, the material gets even more wrinkly, leading to an interesting stippled effect. It tears but is easy to repair with tape.

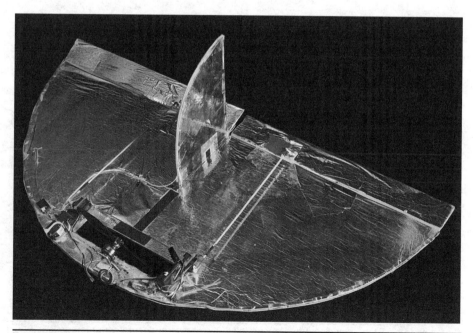

**FIGURE 7-13**  Firefly night flyer with balloon-film covering.

The film is conductive, which may raise issues with radio reception—we always use dual-antenna receivers with this covering. Put the antennas on the top and bottom of the wing to minimize the chance of the material blocking reception. Curiously, the film eventually will wear clear in spots that are handled frequently. It provides no additional stiffness to the foam.

### Assessment of Balloon Film

- *Weight:* 0.10 ounce per square foot, not including adhesive.
- *Technique:* Balloon film applied with spray adhesive.
- *Pluses:* Very lightweight and very reflective covering.
- *Minuses:* Does not stiffen airframe; cannot be made smooth; and may cause radio problems.

## Decorating with Tape

Any sort of colored tape is an excellent decorating material. While the designs tend not to be as enveloping as the earlier designs in this chapter, tape is where most folks find their comfort zone for individualizing their planes. Tape is easy to add to a plane once it is finished. Figure 7-14 shows some of the possibilities based on duct tape, scraps of sign vinyl, and safety tape.

**Figure 7-14**  A plane that has almost all known types of decorative tape applied. Kids like to decorate their planes.

Some considerations on types of tape include

1. *Duct tape.* Heavy, widely available, and often in good colors and interesting patterns from your local hardware store. Be careful of putting too much aft of the CG.
2. *Packing tape.* Excellent material but tends to not be very opaque and can be hard to find in interesting colors.
3. *Masking tape.* Great for accents. Green and blue tape are readily available, but because they are low-stick tapes, they may need clear packing tape on top to keep them in place.

## Conclusion

Decorating planes is very satisfying and well worth doing, but remember that all model airplanes come with an expiration date—especially ours. Always have a flying plane ready to go if you are going to do a big decorating job. It takes longer to decorate than to build usually, so don't let decorating keep you off the field. Chapter 8 addresses decorating for night skies—really fun.

# Make Your Plane Look Good at Night

Night flyers are entities with a super powerful impact on human perception. There has to be some vestigial part of the human brain that harkens back to when we were tiny mammals very worried about what might swoop down from the night skies and eat us. In Manhattan, an alien portal to dimension X could open on Broadway, and people would assume that it was just some marketing stunt—but fly a bit of electroluminescent wire (el-wire) with a wiggly tail 50 feet up, and you introduce jaded New Yorkers to the notion of awe and curiosity. There's nothing like it.

## Overall Considerations

There are lots of different ways to illuminate an aircraft. Illumination has to satisfy two criteria: (1) It must convey to the pilot sufficient information about pitch, yaw, and roll orientation to control the aircraft, and (2) it must impress anyone watching—otherwise, what is the point? Fortunately, satisfaction of point 2 generally takes care of point 1. Choice of illumination is largely determined by the design goal because the illumination effects of the various technologies are so different. Another consideration is use of autostabilization equipment, which makes night (and day) flying much less stressful because the aircraft will right itself when the sticks are released. Chapter 12 covers this technology in detail.

## Major Ways to Illuminate Aircraft

At the night-flying balloon pop competition at the 2012 NEAT (Northeast Electric Aircraft Technology) Fair, all but one competitor used strip light-emitting diodes (LEDs) for illumination. You can already guess that it was the Brooklyn Aerodrome that was the outlier by sticking it out with el-wire, but the visual contrast was remarkable between el-wire and LED lit aircraft. The LEDs could be blindingly bright and very effectively provided wonderful-looking aircraft, but the regularity of the pinpoints of lighting made it very clear that the flying object was engineered. In contrast, the el-wire approach presents as a far more organic shape that cannot help but invoke the biological. The Brooklyn Aerodrome uses both, and Figures 8-1 and 8-2 shows examples of each. Figure 8-3 shows the lighting sources discussed in this chapter.

## Color at Night

Because of how the human eye is constructed, some colors are much more clearly perceived than others at night. In my experience, green really pops at night, with yellow and aqua following closely. Less effective are blues and reds. White light is not particularly effective, but it is fine for 100 feet or less.

## Electroluminescent Wire

The most effective overall solution for night flying is electroluminescent wire (el-wire) in my experience. It looks like thin neon. It is difficult to solder but very flexible and really adds the "alien" to "alien abduction craft." In daylight, el-wire is hardly noticeable, but at night it is a different story. I have flown planes

**FIGURE 8-1**    The Crystal Towel: strip LED edge-lit polycarbonate night flyer.

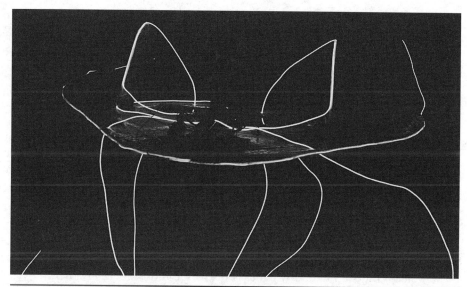

**FIGURE 8-2**    Flying Jelly Fish: el-wire-illuminated night flyer.

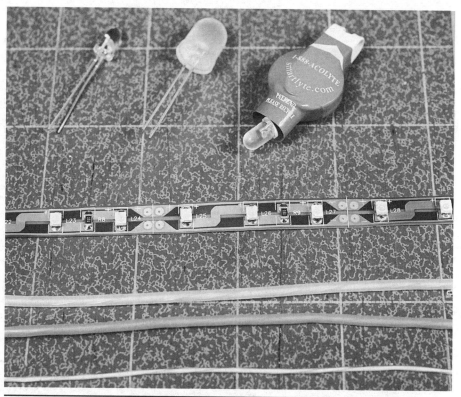

**FIGURE 8-3**    Various sources of light: 3-millimeter LED, 5-millimeter LED, Acolyte battery-powered LED, strip LED, 2.5-millimeter high-brightness el-wire, 2.3-millimeter standard-brightness el-wire, 1.2-millimeter angel-hair el-wire.

at Burning Man where people saw them more than a mile away and ran to find me. The only downside is that soldering el-wire is a considerable hassle. Ready-to-go kits are available, but they will not be able to run off the airplane's power without modification, and modification requires soldering.

## How El-Wire Works

El-wire is made from a phosphorescent coating applied to a core conductor wire with two very thin excitation wires on the outside of the phosphor (Figure 8-4). On application of alternating current (ac), the phosphor coating glows in ultraviolet blue. This color is the source of all el-wire colors. To get other colors, the enclosing sheath is made from various wavelength-sensitive dyes that convert to white, reds, yellows, oranges, and greens.

The only sizes of el-wire we use at Brooklyn Aerodrome are the 1.2- and 2.5-millimeter High Brite from coolight.com at about $1.50 a foot. You will want at least 12 feet, preferably 8 feet of one color and 4 feet of another.

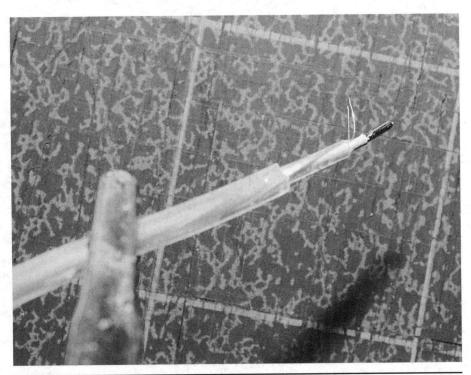

**Figure 8-4**  Properly stripped el-wire ready for soldering. Visible are the outer sheath, inner sheath, two fine excitation wires, phosphor coating, and exposed inner wire.

### Powering El-Wire

El-wire requires an inverter to apply the appropriate ac current. The inverters have a broad range of input voltages and lengths of el-wire they will drive. Length estimates are based on the 2.3-millimeter size. Inverters tolerate overdriving (more wire than rated) better than being underdriven (shorter than specified). Leaving an inverter powered up for long periods of time without a load from el-wire will destroy it. The brightness of the el-wire depends on the voltage of the inverter's output as well as its ac switching rate. An output of 2 kilohertz is considered on the bright side of the distribution at 125 volts. If you touch the bare power coming from an inverter, it will shock you quite solidly, but it is low amperes, so it should not pose much of a danger. Figure 8-5 shows a range of inverters. Coolight.com is our standard vendor for inverters.

### The Easy Way to Get Night Flying

Ready-to-go el-wire kits powered by AA batteries are available for around $25, and that is the easiest way to get night flying. The Flack should be able to handle the extra weight of the inverter and batteries. Get around 12 feet of el-wire, and look at the section on how we route the wire for maximum visibility. A web search should turn up a bunch of options, or check this book's website for the latest information.

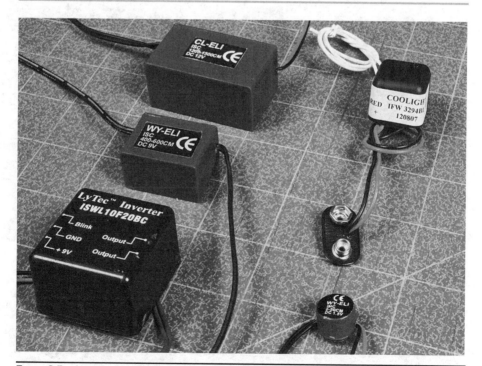

**FIGURE 8-5**  A range of inverters that work with 12-volt input voltages down to 1.5 volts. Lengths that can be driven vary from 15 meters to 80 centimeters.

### IFW 3294BL Inverter

Choice of inverter depends on the length of wire to be illuminated and available power sources. Our standard inverter is the IFW 3294BL from coolight.com for around $9 that will drive 5 to 15 feet of 2.3-millimeter wire at 1.7 kilohertz. This needs around 0.3 ampere at 5 volts of input. The 2.5-millimeter High Brite wire requires more energy, so we use no more than 12 feet total per inverter for the best brightness. As lengths get longer, the inverter drops voltage, and the wire is less bright. The 5-volt input at low amperes allows for very clean installations because power can be drawn from the battery eliminator circuit (BEC) on the speed controller via unused channels on the receiver, as shown in Figure 8-6.

### Soldering an Inverter for BEC Power

The IFW 3294BL inverter arrives with a 9-volt battery connector and bare wires on the el-wire side. *Do not attach* a 9-volt battery to the connector! You will fry

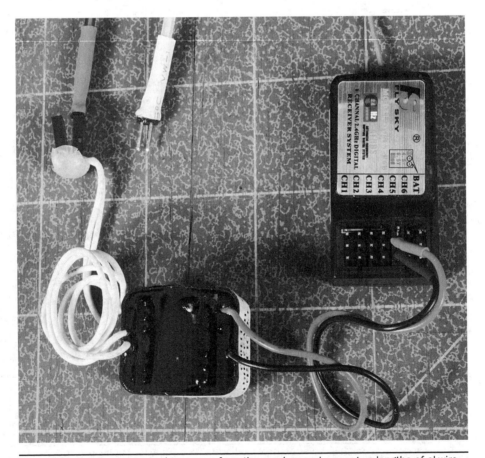

**FIGURE 8-6**    Inverter set up to draw power from the receiver and power two lengths of el-wire.

the inverter. Cut the 9-volt connector off and save it for future 9-volt battery projects. The tools you will need for this include

1. Wire stripper
2. Soldering iron
3. Helping hands
4. Hot-glue gun
5. Heat gun or cigarette lighter
6. Bamboo skewer (useful for manipulating the delicate excitation wires)

Remote-control (RC) receivers are based on 0.10-inch header pins that can be found at various electronics vendors. A website for the desired headers is www.futurlec.com/ConnHead.shtml. Other sites include digikey.com and allelectronics.com. The male versions are described as 0.025-inch square posts on 0.1-inch centers. Female versions can be found by searching on "header" at those sites. There are lots of options.

All that nice, clean 5-volt power coming from the BEC is mighty tempting for powering the inverter. But do visit the "BEC-Powered Lighting" section below to be sure that you are not overwhelming the BEC. The easiest way to access that power is to use a spare RC connector from a stripped servo. There are also two-pin 0.10-inch female header pins that fit the receiver male pins perfectly from electronics suppliers. I don't use the three-pin connectors because the connectors with heat shrink tend to be bigger than the slot in the receiver. These instructions assume that 0.10-inch female headers are used. Instructions for soldering inputs/outputs are given below.

### Adding 0.10-inch Connectors to Inverter Inputs and Outputs

The typical Flack setup for night flying uses one 8-foot and one 4-foot section of el-wire that we wire in parallel. El-wire can be wired up in series, but nearly every illumination job I have done has had parallel implementations. The steps include

1. Strip and tin the black and red wires.
2. Tin a female 0.10-inch header pin.
3. Solder the black and red wires to the female header, and heat-shrink it all up.
4. Strip the white wires and tin them.
5. Using helping hands, solder two female 0.10-inch headers as shown in Figure 8-7.
6. Trim any extra wire on the top.
7. Put blob of hot-melt glue on the wires to insulate them.
8. Slide heat shrink over the bare wires and hot melt. Apply the heat gun.

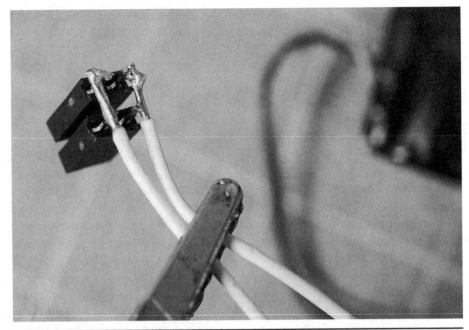

**Figure 8-7**    Two female 0.10-inch header pins soldered to outputs of an inverter.

Never run the inverter without el-wire attached or it will burn out. In my experience it will tolerate brief moments without an el-wire load.

### Soldering El-Wire

El-wire is tricky stuff to work with, but with patience, it can be done. Matt, our test builder for this chapter at Brooklyn Aerodrome, got all the el-wire done, but it was challenging. I highly recommend that you consult this book's website for videos showing the process. Take your time, and keep in mind all the cool stuff you will be able to do if you can master this. The following steps take you through stripping the el-wire and then attaching it to the inverter.

### Stripping the El-Wire

The most difficult part of soldering el-wire is stripping the insulation off without breaking the two very fine wires that wrap the phosphor-coated inner wire. This can be very tough to do with standard wire strippers and may take many tries before you succeed. Do not cut el-wire to length until you have soldered it because you may lose considerable amounts trying to get it properly stripped and soldered. Below are some techniques to try depending on what kinds of equipment you have.

### Option 1: Standard Wire Stripper

The standard wire stripper can work quite well, but it takes skill not to break the fine wires. This works for me about 50 percent of the time. Each failure eats up about ¾ inch of el-wire. Steps include

1. Using a larger stripping size, remove ¾ inch of outer sheath, as shown earlier in Figure 8-4.
2. Using a smaller stripping size, remove ½ inch of inner sheath. It may help to warm the inner sheath with a cigarette lighter. Try to keep the fine wires away from the jaws as you strip to avoid tearing them off.
3. If you get the inner sheath off without breaking the two fine wires, declare success; otherwise, try again. If this goes on too long, try the lighter method below.

### Option 2: Using a Cigarette Lighter to Strip El-Wire

This approach should be done with good ventilation because it really stinks the place up with burned-plastic smell. However, it works well and has a high probability of success.

1. Use a stripper to remove ¾ inch of the outermost sheath.
2. Take a cigarette lighter and burn off the inner insulation so that ½ inch of the small wires are exposed. If you are brave, you can get the inner sheath really hot and just pull off the sooty length, but watch out for burning fingers!
3. Gently clean any soot off the fine wires.

This approach doesn't get much use because option 3 is so effective.

### Option 3: Using the Hobbico 2-in-1 Wire Cutter/Stripper to Strip El-Wire

Figure 8-8 shows the tool that gave me back a month of my life by making stripping el-wire quick and painless. I have a nearly 100 percent success rate with it. It works by gripping the insulation with a cutting set of jaws, and then a pulling set of jaws is actuated with a firm squeeze of the hand. Steps include

1. Use the stripper to take off ¾ inch of the outer sleeve.
2. Reposition so that the inner sleeve will be have ½ inch stripped. Try to align the fine inner wires away from the cutting jaws and strip.
3. Be in awe at how easy that was.

The only weakness of this tool is that the handle will break eventually. I use it only to strip, not cut. I have repaired mine with a bit of reinforcement, epoxy, and a coat hanger to hold it all together.

Once you have the fine wires exposed, it is time to solder.

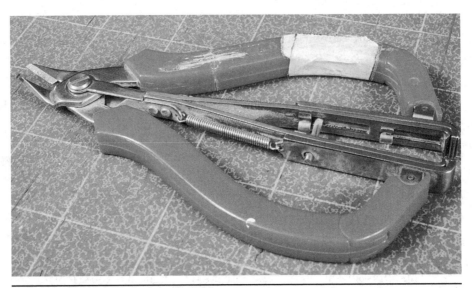

**Figure 8-8**    Hobbico 2-in-1 wire cutter/stripper.

### Soldering the El-Wire to Male Header Pins

Next up is how to connect the el-wire using 0.10-inch header pins, as shown in Figure 8-9. There are lots of other ways to connect el-wire to wire that can be found on the Internet, but I find them slow to execute. The way shown here has served me very well over the years in demanding environments. The steps include

1. Scrape off ¼ inch of the phosphor coating on the inner wire of the el-wire with a knife, wire cutters, or whatever works for you. The wire needs to be completely clean and shiny.
2. Break off a section of male header pin with two pins.
3. Insert the one side into a female connector, and place in the helping hands. The female connector helps the male connector keep its shape when being soldered. This is the same trick I used for soldering the battery connector in Chapter 3.
4. Tin the exposed posts of the male connector.
5. Put el-wire in the opposite jaws of the helping hands.
6. Tease out the two fine conductors from the el-wire, and put them off to the side. I find that a bamboo skewer helps with this. You also should tug gently to verify that the wires are still attached.
7. Tin the core wire.
8. Solder the core wire to one of the header pin posts.
9. Solder the fine wires to the other post.

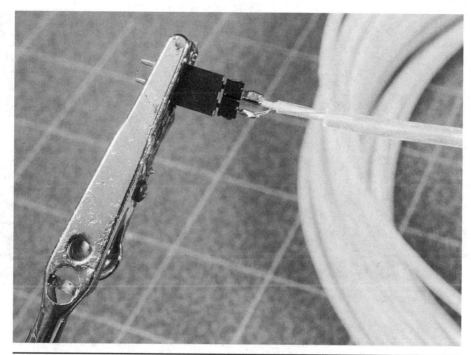

**FIGURE 8-9** El-wire soldered to 0.10-inch male header pins.

10. Inspect the wires carefully to ensure that both fine wires are soldered and that there is not a short circuit between the core wire and the fine wires.
11. Put a blob of hot-melt glue on the solder joints and extend it down to the insulation, as shown in Figure 8-10.
12. Quickly, before the hot-melt-glue cools, slide heat-shrink tubing over the header pins, and squish the glue around a bit.
13. Apply the heat gun to shrink the heat-shrink tubing. Ideally, you should see a little of the glue ooze out the back. This will make for a very strong joint.

Always test the inverter and el-wire before attaching them to an airplane. It is much easier to debug problems on the bench than on the airplane. Steps to vivify the el-wire include

1. Insert the el-wire leads into the female header pins on the output side of the inverter (white wires). Since there is no polarity, there is no way to screw this up. The one mistake you can make, but not a disaster, is to put one el-wire onto one output lead and the other onto the other output lead, thereby not creating a circuit.

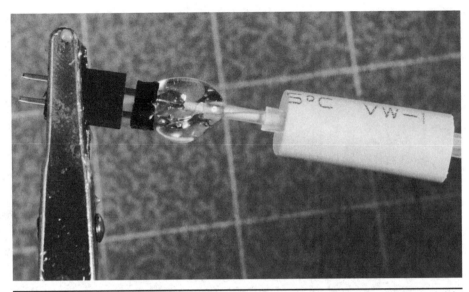

**FIGURE 8-10**    Hot-melt glue applied to a new el-wire joint. The next step is to slide heat-shrink tubing over glue and shrink it all up.

2. Connect the direct-current (dc) power side of the inverter to a spare channel of the receiver. Make sure to get the polarity right, or the inverter may be fried. Brief moments of incorrect polarity do not seem to hurt the inverter, however.
3. Power up the airplane, and the wire should light.

That's it! Time to decorate the airplane.

## Simple Decoration

Make up a 4- and an 8-foot length of High Brite 2.5-millimeter el-wire in contrasting colors. If you only have one color, go ahead with what you have on hand. The suggested placement conveys plenty of orientation without needing color. Figure 8-11 shows a typical installation of the inverter and el-wire. Things to note are that the inverter has zip ties protecting against the direction of crashes, as was done with the receiver and speed control, and the connectors between the inverter and el-wire are well secured with zip ties on both sides of the connection. Also be careful not to make the el-wire turn too tight a corner. The minimum diameter of a turn should be ½ inch, or you risk cracking the phosphor coating, which will short out the entire length of el-wire and keep all el-wire from the inverter from illuminating.

Figures 8-12 and 8-13 show Bill Suroweic's technique of attaching el-wire. It works very well as a way to convey the airplane's orientation and looks great in

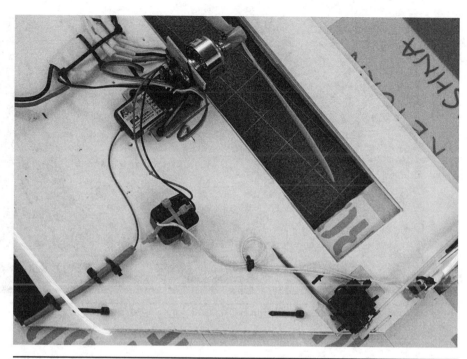

**FIGURE 8-11**    Inverter is powered from the receiver. El-wire starts at the stabilizer.

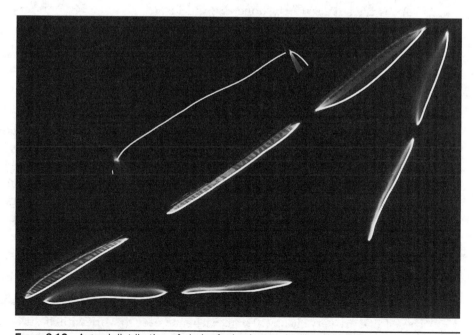

**FIGURE 8-12**    A good distribution of el-wire for beginners developed by Bill.

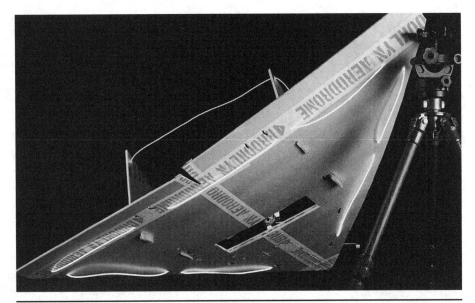

**Figure 8-13**   Better-lit version of the Flack rigged according to Bill.

the sky—the key bit is to have a strand of el-wire go across the tops of the stabilizers. Note that the el-wire is "sewn" through the various surfaces. I prefer this approach for the simpler airplanes because the el-wire stays put better than if tape is used. It also can be done very quickly.

Next up is the world of LEDs.

## LED-Based Approaches

Light-emitting-diode (LED) technology is advancing continually at a brisk pace. The simplest is a single-point LED, followed by strips of LEDs and high-power LEDs that can be used to overdrive optical fiber for beautiful flowing effects. The intense points of light are not as compelling on their own. For example, the Crystal Towel (shown in Figure 8-1) gets much of its visual impact from the reflections in the polycarbonate wing rather than from the LEDs in the edges by themselves. Done correctly, though, LEDs can really make a flying object pop in the sky. Let's start with the simplest approaches.

### Single-Point LEDs

When most people think LEDs, they think of the lights shown in Figure 8-14. These LEDs are very cheap (10 cents each) and easy to work with. If the desired

**FIGURE 8-14**   Navigation lights using single LEDs.

effect is the navigation lights found on full-scale aircraft, then LEDs are an excellent choice. Early Brooklyn Aerodrome night flyers had single-point navigation lights that mirrored full-scale aviation lights for night flying. We have since found that well-articulated designs provide more than enough information about orientation of the airplane for night-flying control.

LEDs run off dc voltage, and they require current-limiting resistors to prevent their burning out. The formula for calculating the required resistor is

Resistance = (power-supply voltage − LED voltage drop)/LED current

Typical values for running off the BEC are 5 volts for power-supply voltage, 2.4 volts for LED voltage drop, and 20 milliamperes of current, yielding a needed resistance of

$$(5-2.4)/20 = 0.13 \text{ ohm}$$

If the calculated resister value is not available, pick the next-higher value. There are many online calculators for calculating the resistor values. You can wire them in series, in which case you add the drop voltages. Two of the preceding LEDs in series would not need a resister at all.

### Battery-Powered LEDs

The simplest version is an LED attached to a watch battery. LEDs are so power-efficient that they will burn for days off a watch battery. They are available commercially at www.wholesaleflowersandsupplies.com/reusable-floral-led-light-III-safety-light.html for less than $1 each, and they turn on and off. There are do-it-yourself (DIY) sources that provide the raw materials as part of the "throwable" culture where magnets are included for urban "decoration." Do a web search for "LED throwies" to go this route.

A major advantage of the watch-battery power source is that the LED stays on no matter what happens to the airplane. I have used them as backup lights that will not turn off if the battery pack is knocked off on a hard landing or crash. This is very useful for airplane recovery if you are flying in large, open dark areas such as Burning Man.

### Strip LEDs

Figure 8-15 shows strip LEDs up close on the Crystal Towel refracting off the polycarbonate flutes in the greenhouse roofing from which it is made. The best prices to be had are from e-Bay or direct import from China, which can be as low as $2 per meter from http://www.aliexpress.com. The good prices start at 5-meter-long rolls, and the range of choices is truly impressive. Strip LEDs can be impossibly bright to look at directly for any period of time, and their

**Figure 8-15**   Strip LEDs reflecting off 6-millimeter polycarbonate.

uniformity makes them look like Christmas lights. The strip lights are only really flexible in one dimension but will tolerate limited twisting and no flexibility in the width of the ribbon.

The only issue around the use of strip LEDs is that they require 12-volt power, drawing 2 to 4 amperes for 5 meters, which, in turn, requires a three-cell lithium-polymer (LiPo) pack to power. The voltage range will be from 12.6 to 9.0 volts, which I have found sufficient to keep the strip LEDs lit. There are a couple of solutions to consider:

1. Have separate three-cell pack with sufficient capacity for a few flights. If the power draw is 2 amperes per hour and flights are 6 minutes, you can expect to draw 200 milliamperes. With some margin, a 300 milliampere-hour three-cell pack could work well. Note that there is no overdischarge protection built into this system, which is one reason that I generally avoid separate power.

2. Convert the airplane to three-cell operation, and draw from the flight battery. The extra voltage will translate into more revolutions per minute for the motor. You may be able to keep the two-cell motor if the prop size is dropped by 1 or 2 inches. Check the motor for overheating. Otherwise, get a lower-kilovolt motor, which turns the prop more slowly per volt supplied. Power from three cells is my preferred route for strip LEDs because there is no danger of overdischarge, and every time I fly, I know that the power to the LEDs is from a battery that should be freshly charged. Even if a discharged battery is used, the motor will shut down before the lights go out.

### High-Power LEDs and Fiberoptics

I was introduced to high-power LEDs by the costume artist BriteLite. I have collaborated with him over the years, and his current bleeding-edge technology is high-power LEDs focused into fiberoptic cable, as shown in Figure 8-16. There is a very high-power LED that has all its light driven into a bundle of fiberoptic cable that leaks light from the sides of the fiberoptic strands as well as the intended route of the ends of the fibers. The results are subtle but stunning.

This setup requires 12-volt dc at 0.5 ampere, and the housings need to be in decent airflow. I have yet to use this setup, but I am really looking forward to the moment of creative clarity when my design will take to the skies. Check this book's website for more information as I learn more about these sources.

**Figure 8-16**   Details of a high-power LED and fiberoptic cable.

## Other Illumination Options

The limiting factors in illumination break down to weight, power needs, robustness, and visual impact. Lighting techniques change all the time, so it make more sense to understand these limitations for evaluation rather than to guess where lighting may go.

### Weight Considerations

The Flack weighs about 15 ounces and can easily carry 4 to 8 ounces of payload. This then becomes the weight budget for any lighting technology. More weight than this will affect flying performance significantly and generally should be avoided for safety reasons, if nothing else. Another factor is where the lighting weight will be with respect to the airplane's CG. The Jelly Fish in Figure 8-2 was initially decorated with el-wire tendrils hanging from the elevon hinge line, which meant that as the tendrils moved, they changed the CG, constantly making the airplane a challenge to fly. It was pointed out by Mike Phillips, whom we met at Burning Man, that the CG would be more constant if the tendrils were concentrated at the CG. The tendrils were moved, and we had a much better flying plane.

Any uniformly lit airframe will have the lighting tend to make the airplane tail-heavy because 75 percent of the wings' surface area is aft of the CG. A heavily decorated Flack will have all nondisplay parts of the lighting (i.e., inverters, extra batteries, etc.) as far forward of the CG as possible to counteract this tendency for tail heaviness. The motor/battery/prop also can be moved forward, which is seen often in our semicircle planes and the bat from Chapter 9.

### BEC-Powered Lighting

The BEC from the speed controller typically supplies between 2 and 3 amperes of power at 5.0 volts, which means that we have between 10 and 15 watts to work with from that source. Servos can draw 1 ampere at stall with much less at rest. With a 2-ampere BEC, we keep the lighting-power draw down to 0.5 ampere, and with a 3-ampere BEC, we will allow 1.5 amperes for lighting. It is easy to see whether the BEC is struggling if the attached lighting is flickering when it should not, particularly when maneuvering hard. If the plane ignores control for a second or two when the lights are flickering, then there is clearly a power issue because the receiver is rebooting on low-voltage brownouts.

If you are dedicated to a BEC-driven solution and need more amperes, then get a separate BEC circuit from companies such as Castle Creations, which offers a 10- and 20-ampere BEC.

### Robustness

I have tried to use halogen bulbs on airplanes to very limited success because the bulbs kept burning out from hard landings and vibration of the plane. Other fails include single LEDs, which tend to get torn off on landings, and glow sticks, which are hard to attach and don't give much illumination. A downside of strip LEDs is that they are hard to repair, and el-wire can get stress fractures that will shut down the entire wire. Keep in mind that the airplane lands on its belly and vibrates constantly when evaluating new technologies.

## Conclusion

Lit planes look great in the sky, are not hard to fly, and the ability to fly at night increases the amount of time that you can fly. In summer, nighttime is one of the better times to fly in parks in Brooklyn. It is also an exciting time to be creating night art because of the innovations that are coming to lighting, and a bit of awe and curiosity is a great gift to a jaded world.

CHAPTER **9**

# Other Shapes

The great thing about inexpensive gear and quick build times is that they encourage experimentation. If a new design completely fails to fly, all I have lost is a few evenings and $5 worth of foam. While my ego may take a beating, my wallet and schedule have not. Another advantage of quick builds is that I can finish the experiment before my enthusiasm runs out. This chapter exists because it is so easy to play.

This chapter gives sufficient information for an experienced Brooklyn Aerodrome hacker to re-create these planes. I address why the airplane was built, design considerations, potential weaknesses, and areas of possible improvement. Some are very difficult to fly; none are hard to build. Have fun, and send me pictures of what you come up with to bible@brooklynaerodrome.com.

## Inorganic Produce

The Banana, Strawberry, and Carrot planes in Figure 9-1 came to be as the result of Canadian artist Andrew de Freitas.

Here's the initial communication:

> I am an artist and filmmaker originally from New Zealand but currently living and working in Montreal. I am in the process of developing a meandering film that weaves a narrative through documentary footage. The starting point for the film is a set of fibreglass fruits hanging from an awning at a grocer's store here in Montreal. By tracing the history of these strange, enlarged objects, the film will take various turns. One turn I would be very interested in taking is toward the Brooklyn Aerodrome. Do you think it would

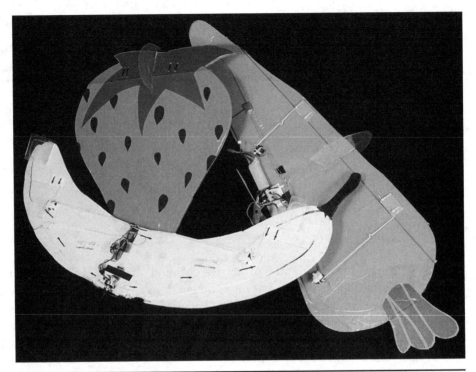

**Figure 9-1**   The planes of a collaboration with Andrew de Freitas.

be possible to make a flying banana or other kind of fruit item? I am eager to know and hope that—if it is possible—we can work together and incorporate this into the film project.

Andrew came to New York, and together we prototyped, built, and shot video over the course of a few days. All the planes were drawn out on blue foam and decorated by Andrew. I then added the flying gear—it was a fun collaboration. You can see the gliders we tested in Chapter 11. Technical details of the aircraft are given below.

## Banana

The Banana reigns supreme in the minds of children as the most popular Brooklyn Aerodrome aircraft ever, other than the Flack. It is a shame that I rarely can fly it at fairs because it needs a lot of room to take off and land. It flies surprisingly well once it has some airspeed. It has the best glide ratio of any of our aircraft. Figures 9-2 and 9-3 show the outline. Note that there is a deck spanning the wing, adding considerable strength.

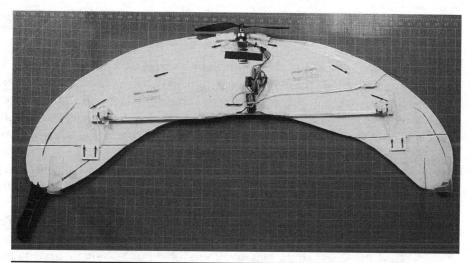

**FIGURE 9-2** Top view of the Banana.

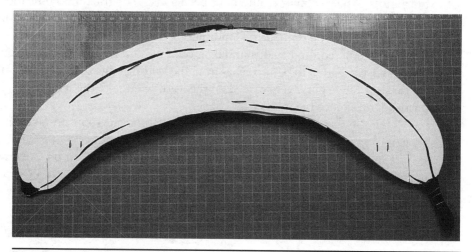

**FIGURE 9-3** Bottom view of the Banana.

Its characteristics are as follows:

- *Wing span:* 46 inches
- *Chord at center:* 8.25 inches
- *Center of gravity (CG):* 4.5 inches from the leading edge
- *Surface area:* Unknown
- *Wing loading:* Unknown
- *Weight:* 581 grams
- *Construction:* Coroplast; blue foam covered with paper gift wrap; food-container lids for stabilizers

- *Flight performance:* Stable; long launch and landing area needed; not particularly acrobatic

Given the odd shape, the control surfaces were a bit of an experiment. They were sized by the "that looks about right" (TLAR) method, and the vertical stabilizers were cut out of clear take-home containers and attached to the elevons, as shown in Figure 9-2. The Banana flies remarkably well without stabilizers at all, but it requires a bit of subtlety to control the adverse yaw that the fixed stabilizers handle automatically.

Note that the servos are very close to the control surface, which made it simpler to set up the elevons, but extensions had to be soldered onto the leads. The outer holes of the servo arms were used to maximize control throw. I often do this with prototypes with unknown flight characteristics to ensure sufficient control authority. These throws are appropriate and fly very well.

The Banana and all the "inorganic produce" planes were set up with tractor propellers to minimize the visual impact of the propulsion system. Prop holes in the middle of these planes would have been a visual distraction. While they are not as safe as our midprop designs, these planes are much quieter and more prone to motor/prop damage on landing.

The Banana required a substantial amount of nose weight to balance on the CG, which was found to be 4.5 inches from the leading edge by experimentation. Discarded tire weights were placed on the leading edge of the wing around the propeller to achieve balance.

Flying the Banana is a pleasure if there is sufficient room to take off and land. It can barely loop and roll, but it does to a reasonable job of gliding, and people love to see it fly. I have flown it at the Figment Festival on Governor's Island in New York over the past few years. Despite being covered in paper, it has held up remarkably well, and there is no reason not to expect to see it for a third year at the Figment Festival.

## Carrot

The Carrot was a particular challenge because it didn't present an obvious direction of flight. Flying with the sharp end forward is the most obvious approach, but very low aspect ratio wings can be tricky to fly, and I was not sure of how to get the control surfaces functional. Pitch control should not be a problem, but given the narrow wing, roll control might be an issue. Flying it blunt end first could work but was even less appealing from a control-surface standpoint.

The decision was made to fly it edgewise, which had the consequence that the Carrot was asymmetric, with one short, stubby wing and one long, thin wing, as shown in Figures 9-4 and 9-5. Getting this arrangement to work required determining where to split the wing in half. The details of how this

**Figure 9-4** Top view of the Carrot, one of the few asymmetric aircraft at the Brooklyn Aerodrome.

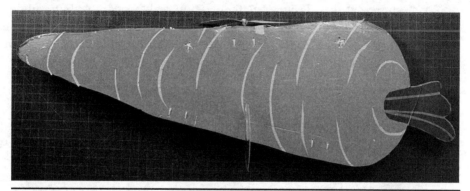

**Figure 9-5** Bottom view of the Carrot, quite pretty for a vegetable that few people care for.

was done are given in Chapter 10, but the short story is that the short, stubby wing is not as efficient as the long, thin wing in generating lift, so it requires more surface area. This, in turn, forces the center of the wing 2½ inches to the right. There is a large clear food container top acting as a stabilizer between the elevons, which was adequate, but still very close to the CG, which limited its effectiveness. Twin stabilizers may have been a better choice. As with the Banana, it needed a substantial amount of lead weight to achieve the 3-inch CG as measured from the leading edge at the motor.

Flying the Carrot was quite a challenge. It had a bad tendency to tip stall, so it was vital to keep the airspeed up (a *tip stall* is when one wing stops generating lift, and the plane spirals toward the ground). It turned better to the right than the left because the high-aspect-ratio wing (long and skinny) did a better job of generating lift on the inside of the turn than the short, stubby wing. It was not a very fun plane to fly. It probably could be made to fly better with more work, but once we were done with the movie shoot, it never became a priority.

- *Wing span:* 48 inches
- *Chord at center:* 12 inches
- *CG:* 3 inches from the leading edge at the motor
- *Surface area:* 416 square inches
- *Weight:* 20 ounces
- *Construction:* Coroplast; blue foam covered with paper gift wrap; food container lids for stabilizers
- *Flight performance:* Nasty tip stall; turns better to the right; a handful to fly

### Strawberry

The Strawberry was the conceptually easiest plane to design because it was quite close to being a delta wing. It was the only one of the "inorganic produce" collection that was designed without a full-scale prototype, but a glider was used to test its asymmetric flight properties that ultimately were not used in the design. It uses classic construction with a deck, tractor prop, and standard elevon layout and a clear stabilizer, as shown in Figures 9-6 and 9-7.

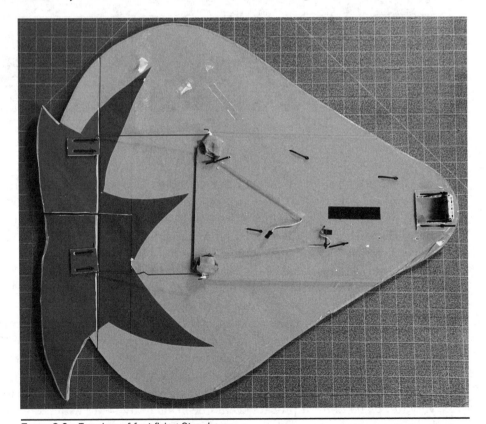

**FIGURE 9-6**    Top view of fast-flying Strawberry.

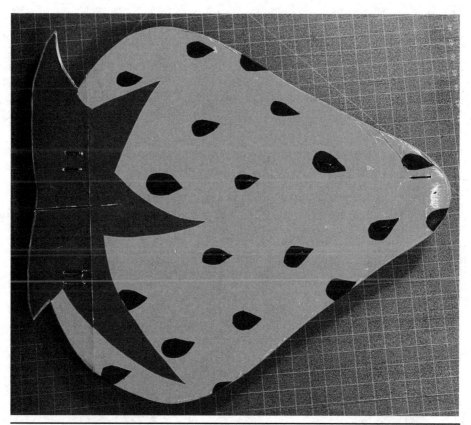

**Figure 9-7**  Bottom view of the Strawberry.

Because it is the smallest of the three planes, it had the highest wing loading and was quite aerobatic. It did not glide very well, though, and it had to be powered until just before landing. It did not stall gracefully and needed to be flown fairly fast.

- *Wing span:* 23 inches
- *Chord at center:* 26 inches
- *CG:* 10 inches from the leading edge
- *Surface area:* Unknown
- *Weight:* 16 ounces
- *Construction:* Coroplast; blue foam covered with paper gift wrap; food container lids for stabilizers
- *Flight performance:* Fast airplane with very poor glide ratio; quite aerobatic

Of the "inorganic produce" series, only the Banana sees continued use because it flies so well and amuses onlookers so much. The Carrot just doesn't seem to generate the same enthusiasm. Next up is the Flying Heart.

## Flying Heart

The Flying Heart was created for the Escape2NY Music Festival. The festival was done by the same folks who do Secret Garden Party in England, which is a Victorian-themed event, so something *Alice in Wonderland*–based was desired. What better than a flying heart that ended up looking like a flying spade (Figure 9-8).

### SketchUp Is Your Friend

The basics of the design were done in Google SketchUp, which provided some crucial advantages over the free-hand approach used with "inorganic produce." I used SketchUp to do the following:

**Figure 9-8**  Top view of the Flying Heart.

1. I explored different versions of the basic heart shape. SketchUp allowed me to experiment with prop placement, width/length, and fitting inside the travel container.
2. The surface-area calculation in SketchUp allowed me to maintain important design parameters, including
   a. Keeping the wing area of the Heart similar to that of the Flack. With similar wing loading, I was more confident that the airplane would fly with standard Flack gear.
   b. Mirroring the elevon area of the Flack.
   c. Determining the likely CG of the Heart. It was at the line that divides the wing into 25 and 75 percent of the surface area.
3. I produced full-scale plans to guide cutting the foam on a series of 8½- × 11-inch sheets. The full design was printed out onto 20 sheets. Joining them was quite time-consuming. To make joining and aligning the sheets easier, I added a chain-link-fence pattern to the wing, as done with the Manta Ray later in this chapter.

Once the Heart was cut out of the foam and the tail removed from the center, I began to flex the foam to see what kinds of reinforcement would be needed. I built two versions of the heart, one with purple balloon film and the other with sign vinyl. Since neither of these materials adds much stiffness, a Coroplast deck was added from stripes 2 through 5 back from the nose. The motor mount also was on Coroplast at the ninth stripe back. Note the hangers that have been added between the fore and aft Coroplast sections to add stiffness. If I were to build the Flying Heart again, I would extend the deck to be continuous with flutes going fore to aft.

The placement of the motor raised the issue of where the electronic speed controller (ESC) would go. Because of CG considerations, the battery would have to go as far forward as possible. Two options exist:

1. Bring the battery power to an aft-mounted ESC and receiver, which would allow very close placement of servos to the elevons.
2. Keep the ESC close to the battery, and run long motor leads and long control rods.

A largely unknown aspect of ESC design is that ESCs are not tolerant of long battery leads because of voltage induction spikes that occur when the ESC is switching on and off at a very high rate. So option 2 makes sense. The ESC can be modified with capacitors if there is no other solution.

A novel feature is that the tail is mounted on a Coroplast hinge that allows it to kick up on landing. It works very well.

The Flying Heart remains a huge success for the Brooklyn Aerodrome. The striped version is visually striking and flies like it is on rails. It is not very acrobatic, but it is a great cruiser that responds well to difficult flying conditions. It would likely make a fun night flyer as well.

*Wing span:* 32 inches
*Length:* 41 inches
*CG:* 10.5 inches from the elevon hinge
*Surface area:* Unknown
*Weight:* 605 grams
*Construction:* Coroplast; blue foam covered with balloon film or sign vinyl; bamboo skewers
*Flight performance:* Great flying airplane that cruises very well; has nicely controllable glide and climbs well

## Travel Flack

The Travel Flack is meant to be a maximally portable version of the Flack. The box in Figure 9-9 contains a Flack and a charger and transmitter in a 15- × 16- × 3-inch case with a shoulder strap. It has flown as far south as Savannah, Georgia, as far west as the Black Rock Desert in Nevada, up to Warren, Vermont, in a northern direction, and as far east as East Hampton, New York. It is also a night flyer and has been taken out by a snowball in Alta, Utah. Visit this book's website for more detail about this design.

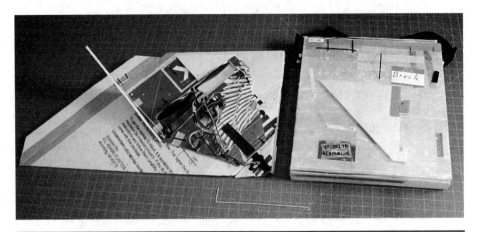

**FIGURE 9-9**    Travel Flack halfway folded up.

## Bat

The Bat (Figure 9-10) is another SketchUp-designed plane that would have been very difficult to create without the flexibility of the tool. Like the Flying Heart, the Bat had to fit in a standard-sized shipping container.

The Bat did not go through a prototyping phase because I was extremely confident that it would fly given the surface area of the wings and their shape. What was more challenging with the bat was determining how to get the predicted CG given the size of the airframe and have it fit in the same amount of space as the Flack.

The interesting features of the airplane include the following:

1. Stiffening of the 48-inch span was achieved with another layer of blue foam glued in the back and zip-tied on the front edge, as shown later in Figure 11-4.
2. The covering was balloon plastic that added no additional stiffness.
3. The stabilizer is just another copy of the wingtip cut at 11 inches from the tip.
4. Electroluminescent wire (el-wire) was loosely attached to the elevons with enough slack to allow for full up and down movement.
5. The plane used an extra battery pack to balance the CG. The extra weight did not seem to affect flight performance very much. If I had been thinking, I would have driven the el-wire off the extra battery.

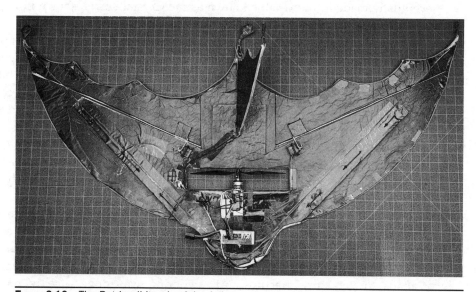

**Figure 9-10**   The Bat in all its aluminized glory.

The Bat is the most relaxing night flyer I have ever flown. It flies like it is on rails, is extremely stable, and very clearly presents orientation information that significantly reduces pilot workload when flying. It does take a while to lose altitude with a fairly good glide ratio. It was flown a great deal at Burning Man 2011 and at Escape2NY.

- *Wing span:* 48 inches
- *Length:* 24 inches
- *CG:* 9 inches from the nose
- *Surface area:* Unknown
- *Weight:* 630 grams
- *Construction:* Blue foam covered with balloon film; el-wire edge illumination
- *Flight performance:* Great flying airplane that cruises very well; has nicely controllable glide and climbs well; look great in the air; most relaxing night flyer ever

## Crystal Towel

A pilot walks into a bar carrying a Flack after a successful flight test. Some guy sitting next to him says, "Wow! What is that?" In the ensuing conversation (sorry, no punch line), it turns out that Pierre, who does sets for movies and plays, had a bunch of material that he was going to throw out. The result of that conversation was acquisition of polycarbonate greenhouse roofing that eventually made it into what is known as the Crystal Towel, shown in Figure 9-11. It is clearly a creature of the night, as shown in Figures 1-4 and 8-1.

The wing and stabilizers are made of 6-millimeter polycarbonate with pink strip LEDs illuminating from the edges of the wing and stabilizers. The plane is built to the same dimensions as the Flack with the elevons extended a bit to make the airplane a perfect triangle. The greatest refractive effect is achieved by placing the LEDs parallel to the flutes, which motivated splitting the wing down the middle. That, in turn, led to needing to add a stiffening spar across the middle of the wing.

The polycarbonate is very difficult to cut with a knife. I did not use a saw because I didn't want polycarbonate dust in the fluting. It was probably not worth the trouble.

To run power to the LED strips, I created a system of traces using stained-glass copper edging tape. The tape is very useful for low amperage (1 ampere or less power routing on rigid airframes), which is sufficient for servos and various lighting applications. The copper tape is not good for airframes that flex because it tends to get stress fractures. All soldering was done directly on the polycarbonate, but it has to be done quickly so as not to burn the plastic.

**FIGURE 9-11**    Crystal Towel in its most unassuming state. Crash damage is visible on the nose.

Fully built, the Crystal Towel is quite a bit heavier than the standard Flack. The polycarbonate weighs 4.2 ounces per square foot compared with 0.8 ounce per square foot for raw blue foam. This means that the design will be tail-heavy with standard components because most of the mass of the airframe is aft of the CG. Rather than use lead for nose weight, I opted for a bigger battery and bigger motor, which required a bigger speed controller. Table 9-1 presents a comparison of the different components and their weights.

In designing the Crystal Towel, I did not cut the prop hole until I could test whether the increased weight of the battery and speed controller could offset the tail weight added by the polycarbonate. The motor and propeller are so close to the CG that it was not a consideration. It turned out that the plane balanced with the battery at the very end at 12 inches from the nose—the three-cell A123 is 2.5 times heavier than the standard lithium-polymer (LiPo) pack. All up, the Crystal Towel was just a bit over 1 kilogram, or 2.2 pounds.

How did it fly? Flying the Crystal Towel is terrifying both in the daytime or at night because it is very hard to maintain orientation. During the day, the plane is mostly transparent and disappears into the sky, and at night it presents an outline that acts like the Necker Cube visual illusion. Figure 9-12 shows the classic visual illusion, where the marked vertex snaps back and forth between being the nearest edge and the furthest away edge. The Crystal Towel has a similar effect, where the plane looks like it is flying away from the pilot with the marked vertex furthest away from the pilot, which switches to the marked

TABLE 9-1  Comparative Materials of Crystal Towel and Flack

| Parts | Material | Weight Grams | Weight Ounces |
|---|---|---|---|
| **Flack** | | | |
| Stabs | 1/4" blue foam 0.8 oz. sq. ft. | 10 | 0.4 |
| Airframe | 1/4" blue foam 0.8 oz. sq. ft. | 83 | 2.9 |
| Battery | 2 cell 1800 Mah LiPo | 97 | 3.4 |
| Motor | 2208 | 42 | 1.5 |
| Speed controller | Generic 18 amp 2 am BEC | 10 | 0.4 |
| Prop | GWS 9x4.7 | 5 | 0.2 |
| | | 246 | 8.7 |
| **Xtal Towel** | | | |
| Stabs | 6mm polycarbonate 4.2 oz. | 53 | 1.9 |
| Airframe | 6mm polycarbonate 4.2 oz. | 433 | 15.3 |
| Battery | 3 cell A123 LiFe | 254 | 8.9 |
| Motor | Uberall Nipply black | 91 | 3.2 |
| Speed controller | Jeti 33 amp | 41 | 1.4 |
| Prop | APC 10x4.7 | 12 | 0.4 |
| | | 883 | 31.1 |

vertex closest to the pilot with the plane flying in the opposite direction at the pilot!

Because controls are reversed when the plane is flying at the pilot, very different inputs are required. If you are interested in building this aircraft, I strongly suggest that you use an autostabilization system to prevent these confusions from crashing the plane. One system is discussed in Chapter 12.

Additionally, this airplane is heavy, flies fast, and is very rigid. It could really hurt whatever it hits. This is not an airplane that should be flown where there is any chance of hitting things that go ouch. That said, it flies well and handles wind with confidence. You can see it fly at http://youtu.be/ZoUTLGYXee8.

- *Wing span:* 42 inches
- *Chord at center:* 23 inches
- *CG:* 12 inches from the leading edge
- *Surface area:* 475 square inches
- *Weight:* 1,009 grams
- *Construction:* 6-millimeter polycarbonate greenhouse roofing; edge lit with strip LED; more powerful motor; bigger battery

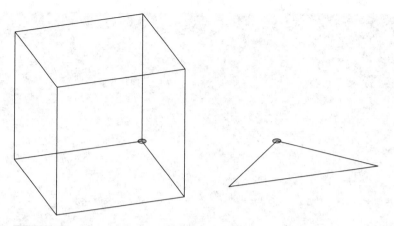

FIGURE 9-12    Visual illusion experienced when flying the Crystal Towel that is similar to the Necker Cube illusion. The marked vertex on both images can appear to be the closest vertex to the viewer or the farthest. When flying, this confusion can be disastrous.

- *Flight performance:* Fast airplane that handles wind well; very difficult to maintain accurate perception; autostablization system highly recommended

## Semi-Circle

The Semi-Circle (Figure 9-13) was an attempt to minimize wing loading by maximizing wing area while keeping the weight down. It was meant for indoor flying at a costume ball. The blue foam sheets are $48 \times 24$ inches, so the decision was made to try a semicircle. The major concern with this design is getting enough weight to the front of the airframe so that the CG is correct. The 25 percent chord point on the wing is 9 inches back from the nose, which is not very much room to establish balance. Add in that there is more material behind the CG with the extra wing area (274 extra square inches over the Flack with a shorter lever arm to balance it), which means that nose weight may be required to balance things out. This, in turn, compromises the goal of having very light wing loading. I achieved a wing loading of 3.4 ounces per square foot, whereas the Flack has a wing loading of 4.0 ounces per square foot. Given that flying speed drops as the square root of mass, there was no appreciable reduction in flying speed.

It is clear that the prop hole has been moved as far forward as possible while still affording some protection from the prop. Carbon-fiber push rods were used to minimize aft-of-CG weight, and decoration was limited to tissue on one side with adhesive film on the other. The decision was made to reverse-mount the motor and have the battery aft of the prop hole for aesthetic reasons (the eye had

**Figure 9-13**    Semi-Circle flying wing with one eye. This was an experiment in light wing loading that failed.

to go somewhere). This was not quite enough nose weight, so two AA batteries were taped as far forward on the nose to balance the plane.

While I don't do this anymore, I was jammed for time and conducted the first test flights in the park on the way to the ball. This is not recommended. While the plane was flown twice at the event, it was curtailed by lack of confidence in the flying qualities of the aircraft. However, it looked great.

The Semi-Circle could stand more development effort. The elevons could be bigger, and more effort could be put into optimizing component placement for CG. To cut flight speed in half would require a four-times reduction in wing loading, so perhaps a very different approach is called for.

- *Wing span:* 48 inches
- *Chord at center:* 24 inches
- *CG:* 9 inches from the leading edge
- *Wing area:* 894 square inches
- *Weight:* 590 grams
- *Wing loading:* 3.4 ounces per square foot
- *Construction:* Blue foam covered in adhesive plastic film on one side and tissue paper on other side; el-wire night-flying decoration added
- *Flight performance:* Flies in a "floaty" way that allows for indoor flying; more work is needed on design to optimize performance; perhaps bigger elevons

# Manta Ray

The Jelly Fish (see Chapter 8) was a success, so more sea creatures flying out of context always seems like a good idea to keep pushing—behold the Manta Ray shown earlier in Figure 1-3. In addition, the Figment Festival was coming, and we could not help but put on a show. It is worth going into the design process a bit with the Manta Ray because it represented explorations both artistically and on the engineering front.

But first, what did we create? The Manta Ray is an animatronic approximation of an actual manta ray. The fins go up and down in a somewhat lifelike fashion in the same way Disney's Country Bear Jamboree emulates bearness, but our Manta does not attempt the banjo or sing. The wing-tip movement is double action, involving two servos per wing, with directional control provided by two elevon servos that work like those on any other aircraft in this book. The tail wags in very un-manta-like ways, but it adds to the visual impact. The Manta Ray looks great in the air. Links to videos are available on this book's website.

## Design

The call for a Manta Ray came from the art department at Brooklyn Aerodrome, and the aero group had to pass judgment as to feasibility. What makes a design feasible? Look to Chapter 10 on making your own shapes fly for a process that increases the chances of success. In this case, we thought the shape was flyable without glider testing, so the issue was creation of a proper Manta Ray shape. Scouring the Internet for the general idea was very useful, and in the end, we had a two-dimensional design of our own making in SketchUp, as shown in Figure 9-14.

Again, we had constraints that the Manta Ray had to be able to fit in a standard Flack shape because all the gear had to be carried in a highly modified baby jogger for transport to the Figment Festival via ferry. Welcome to modern urban flying!

## Evolving the Moving Fins

Almost all our aircraft have a very two-dimensional and static quality to them that makes them less visually salient. The goal of the Manta Ray was to generate movement that would catch the eye. We also wanted the plane to look alive. Figure 9-15 shows the prototype Manta Ray with single movable fins and the production version next to it. The prototype flew well, and the fin movement had little effect on the flying characteristics. The visual impact, however, was less than ideal because the servo throw really didn't move the fins that much

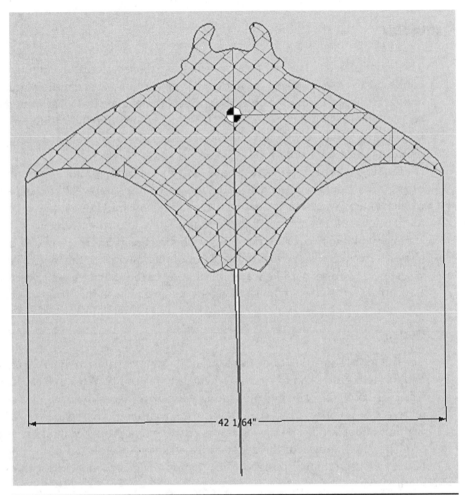

42 1/64"

**FIGURE 9-14** Manta Ray as designed in SketchUp. The chain-link-fence pattern helps to align 8- × 10-inch sheets of paper for scale plans.

with the 45 degrees of throw on each side of neutral, the fins were small, and the single hinge point didn't look particularly natural.

The solution that I went with in the end involved double hinging the fins while retaining the same basic servo control mechanism as shown. This innovation was done on the final version with confidence that it would likely work out because the prototype flew well. In any case, the fins didn't necessarily need to move if some flying issue arose.

There was one potential point of serious failure: Would the servos be strong enough to handle the forces on them? The outer servo would be fine, but the inner servo needed to move a fairly large surface through edge-on flow, and the surface weighed a few ounces. I had no solid idea what the forces would be on

**FIGURE 9-15**   Prototype Manta Ray fin next to production version that has more articulation because of double servo action.

the inner servo. The servo issue was of sufficient concern that I did spend 15 minutes looking for some nice strong metal-gear servos for the inner fins but had no luck, so I went with the 16-ounce torque HXT 900 I had on hand.

To run power to the servos, I used stained-glass copper tape because it looked great, and it probably didn't compromise the power getting to the servos too much. In the end, each servo was independently controlled, meaning that four channels controlled the fins and three channels controlled the throttle and elevons. The decision to have each fin individually controlled was to provide maximum flexibility in programming how it flew and to allow each fin segment to be trimmed independently. Servos shared positive and negative traces, with the signal channel being independent.

## The Radio

When experimenting with "weird configuration" aircraft, it is very useful to have an easily programmed radio with lots of channels to play with. The Brooklyn Aerodrome radio of choice for these situations is the Multiplex P4000

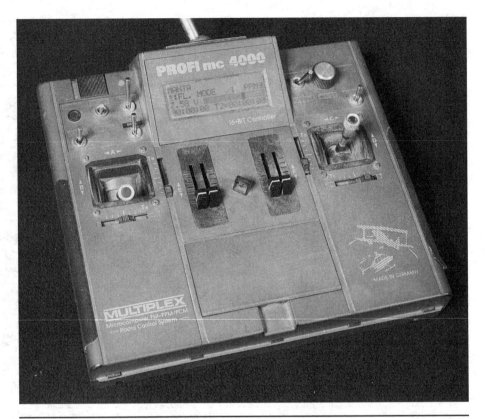

**FIGURE 9-16**   Multiplex P4000 radio.

shown in Figure 9-16. It is no longer being made but can be found used in good condition for around $500. Why the need for the P4000 in this situation? The Manta Ray uses seven channels with some odd properties. Three of the channels are the standard elevon mixing and throttle control. The remaining four were programmed to move with rudder commands (left stick, left to right movement) so that I could control the flapping behavior. The P4000 programming made it trivially easy to assign channels 4 through 7 to the rudder stick, center all the servos, and linearly increase up elevator when the fins were up or down to compensate for a minor but annoying trim change as the fins moved. Almost any seven-channel or up computer radio should be able to do the programming, but in my experience, most software in other radios gets cumbersome when straying outside traditional uses of channels for aircraft.

### The Build

Once the moving fins were sorted out, the rest of the build was standard. The painful process of aligning and joining many sheets of paper for full-size plans was made much easier by applying a regular texture to the image—in this case a chain-link-fence pattern. The flapping tail was achieved by segmenting the foam tail every 4 inches with a tape hinge. As with all the hinges on planes, it is important to use packing tape as the hinge material rather than covering material because coverings tend to get fatigue breaks. All control rods were kept as short as possible to maximize stiffness. Standard Flack motor, speed controller, battery, and props were used.

### Flying

The Manta Ray flies well in wind and has considerable maneuverability and ample power. The movable fins were a bit awkward to use initially, but after a few flights, I found myself going into tight turns with the fins up and exiting with the fins down because the Manta Ray felt snappier that way. It certainly also looked good that way. It is less aerobatic than a Flack, but it climbs quickly and has the same miserable glide performance, meaning that small landing areas are easily managed.

- *Wing span:* 42 inches
- *Chord at center:* 21 inches
- *CG:* 6.5 inches from nose between eyes
- *Wing area:* 413 square inches
- *Weight:* 20 ounces
- *Wing loading:* 7 ounces per square foot
- *Construction:* Blue foam covered in blue adhesive plastic film; no deck
- *Flight performance:* Solid flyer with dramatic look in the air

## 3D Banana

The three-dimensional (3D) Banana addresses a few unsatisfied desires at the Brooklyn Aerodrome. First is that our day fliers are very flat and disappear in many flight orientations that interfere with the visual impact of the design. Second, the 2D flat-plate designs are very limited by what shapes can be flown because both lift must be generated and the shape must be controllable. Third and last, we wanted to represent objects as they actually are in three dimensions, not reduce them to a two-dimensional field. Figure 9-17 shows the 3D Banana.

The 3D Banana is a collaboration with an artist (Chris "Kit " Niederer) who took a 3D model of a Banana in Sketchup and applied a plugin called *Slice Modeler* to create a series of flat plates for the full 3D design shown in Figure 9-18. The resulting aircraft is technically a triplane.

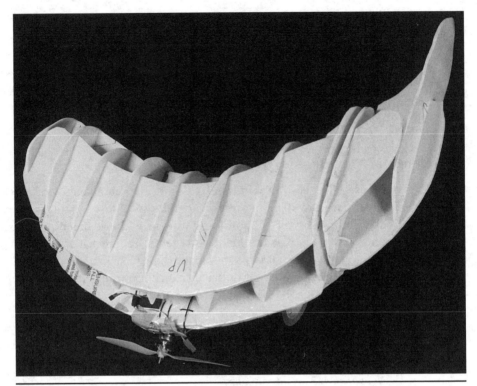

**FIGURE 9-17**    Triplane Banana.

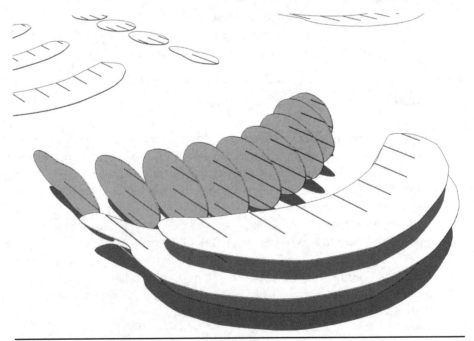

**FIGURE 9-18**    Full-on 3D Banana.

The major challenge of the build was figuring out how to control the shape. The elevons on the two-dimensional Banana were going to be tricky with the triplane, so an alternative approach was taken that mounted the tips of the banana on servos for full flying elevons, as shown in Figure 9-19—this is maybe the first ever example of such a control surface on a flying wing and almost certainly the first on a triplane flying wing. I think it is safe to say this is the first banana-shaped flying wing with full flying elevons, so our place in history is assured.

The 3D Banana is still in development. It has flown a few times and crashed a few because of programming errors that had roll control reversed. It looks to be a draggier version of the two-dimensional Banana the chapter started with, so it could use a more powerful motor.

- *Wing span:* 37 inches
- *Chord at center:* 12 inches
- *CG:* 4½ inches back from the motor—still being determined
- *Wing area:* Unknown
- *Weight:* 22 ounces
- *Wing loading:* Unknown
- *Construction:* Blue foam with Coroplast reinforcement
- *Flight performance:* Flies; needs more power

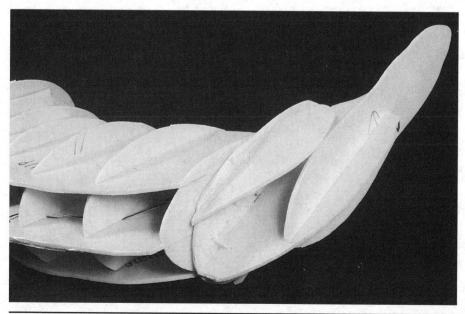

**Figure 9-19**  First ever full-flying banana elevons!

## Conclusion

This chapter laid out the space of possibilities that we have considered here at the Brooklyn Aerodrome. Some lessons learned were that it is really fun to collaborate with artists—if it weren't for these collaborations, I would still be making pure delta wings and hanging wriggly bits of el-wire off of them. Next up is a brief chapter on aerodynamics for the hacker.

# CHAPTER 10

# Aerodynamics for the Hacker

A bit of theory helps with hacking the skies. This chapter covers the major parts of why airplanes fly and how they are controlled.

## Lift

Lots of explanations of how wings generate lift are like a six-year-old's conception of where babies come from—wildly inaccurate but somehow plausible. Below is another wrong explanation, but it gets the ball rolling until you enroll in a proper fluid dynamics class. So here it goes . . .

### Conditions for Lift: Positive Aerodynamic Angle of Attack and Airflow

Take a piece of cardboard and spin it around or put your hand out a car window and notice that upward force happens when the front of your cardboard or hand is higher than the rear with respect to the airstream flow. Compare that to when your cardboard or hand is aligned with the airstream and generates no upward force—this is a neutral aerodynamic angle of attack. When you have a positive aerodynamic angle of attack, there is an upward force or lift. This ought to be obvious by experience.

### Lift, Part 1

Now think about how you are generating the upward force. There is an obvious simple component, which is shown in Figure 10-1, that has air molecules bouncing off the bottom of the cardboard—it is increasing the pressure on the bottom of the wing. You can just think of Mr. Air Molecule as bouncing off the

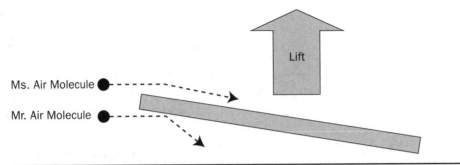

Figure 10-1    The two basic components of lift.

cardboard downwards, which by Newton's Third Law will impart an equal upward motion to the cardboard. For the sake of completeness, you should realize that the cardboard is also pushing Mr. Air Molecule forward a bit, which translates into a backwards force on the cardboard. This is drag. Drag resists the forward progress of the cardboard.

## Lift, Part 2

With all those air molecules ricocheting off the bottom, what happens to the air molecules that just make it over the top of the cardboard? Well, Ms. Air Molecule no longer has Mr. Air Molecule next to her because they were separated by the cardboard. In fact, there is nothing there (a vacuum) unless Ms. Air Molecule moves. That vacuum will draw her in and the other molecules around her down, which will (1) accelerate them and (2) retain less pressure because the same number of molecules will be spread out over a larger volume. My 10-year-old niece Annika observed that Ms. Air Molecule was accelerating like she was on a roller coaster with vacuum replacing gravity. This reduction in pressure is the second component of lift and is called *Bernoulli's principle*. That's it, folks. There are lots of other factors that can come into play like curved airfoils but it is all doing the same basic thing. Send hate mail to crazyLiftTheory@ brooklynaerodrome.com.

## Stall

Stall is what happens when the wing loses the Bernoulli component of lift, which then leads to the airplane no longer having sufficient lift to support it and it falls. The classic version of a stall occurs when the angle of attack is increased so much that the nice vacuum bubble detaches from the top of the wing. There are lots of ways that the bubble can be detached. Well-designed airplanes react to stall by dropping the nose automatically to aid in reducing the angle of attack and increasing airspeed.

## Pitch, Roll, and Yaw

Airplanes can move in three dimensions, and they come with names. Figure 10-2 shows the definitions of *pitch* (nose up/tail down or opposite), *roll* (left wing up/right wing down or opposite), and *yaw* (nose swinging to the left/tail to the right or opposite).

## Center of Gravity (CG)

The CG is simply the single point where the aircraft can be suspended in any orientation and it will remain there. Another way to think about it is as the point where the weight of the airplane balances in pitch, roll, and yaw as shown in Figure 10-2. It is more generally known as the *center of mass* or *barycenter*, but talking that way will just confuse folks at the field. On plans, it is marked with a crossed circle. When you see the CG mark on plans, it means that the airplane needs to balance on that point to fly well, which dictates how components and additional weights are distributed on the aircraft. Rarely is there an issue with CG in dimensions other than fore/aft, but see the Carrot plane, where left/right CG is important, in Figure 10-3.

Why we care about the notion of CG is stability. Generally speaking, lift is pretty easy to generate. Where things get tricky is in controlling that lift so that the aircraft is stable and can maneuver.

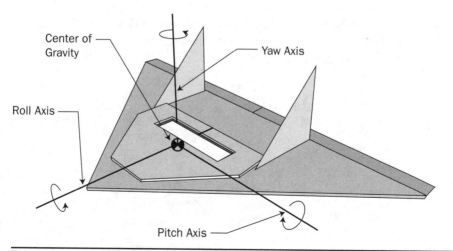

Center of
Gravity

Yaw Axis

Roll Axis

Pitch Axis

**FIGURE 10-2**   Pitch, roll, and yaw explained.

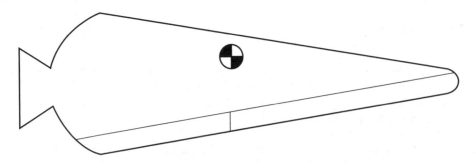

**Figure 10-3**  Asymmetric shapes will require determination of left/right CG as well as standard fore/aft CG. Top/bottom CG can be an issue as well, but it is unlikely with flat-plate designs.

### Yaw Stability

A weather vane's only degree of freedom is yaw. The pivot point functions just like the CG, but it is determined by a bearing. The weather vane in Figure 10-4 is not going to work very well because there is equal area in front of the bearing and aft of the bearing.

The wind is going to spin it around. If this is not clear, go ahead and grab a coat hanger, some cardboard, and a fan and experiment for yourself. How can this be fixed? Two options present themselves immediately:

1. Increase the surface area on one side of the bearing.
2. Move the bearing more to one side.

Figure 10-5 shows each approach. But note that they are essentially doing the same thing, increasing surface area on one side of the bearing and decreasing it on the other. Figure 10-6 shows a side view of the Flack with the bearing at the CG. With those huge stabilizers, you can see that this design is going to point into the wind without any problem.

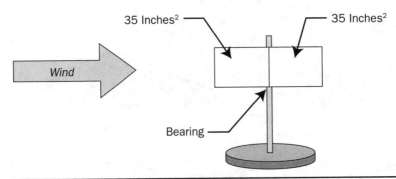

**Figure 10-4**  Weather vane stability demonstration.

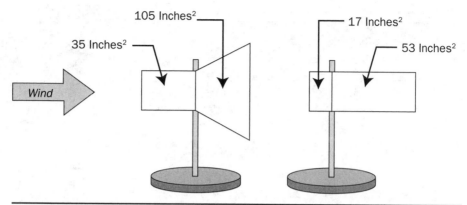

**FIGURE 10-5**  Two solutions to stabilizing the weather vane from Figure 10-4 that amount to the same solution—get increased surface area aft of the hinge point.

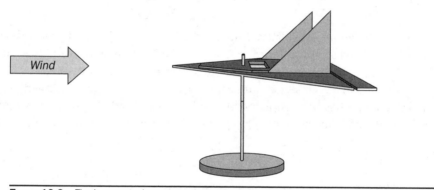

**FIGURE 10-6**  Flack mounted as a weather vane at the design CG.

### Static Stability

The Flack relies entirely on what is called *static stability for yaw*, which is provided by the big stabilizers automatically, just as a weather vane gets automatic corrections from the greater surface area aft of the CG.

### Pitch Stability

Pitch stability is more complicated than yaw stability because lift is being generated in addition to active control surfaces. But each complication will be introduced in turn. Consider the Flack in a straight-down dive as in Figure 10-7. At this moment, the wing is generating no lift, and the airplane is stable in yaw because of the greater stabilizer area behind the CG. Remember that the CG is the balance point of the airplane in all orientations.

Consider what is happening in pitch. We can just pretend that the surface area of the wing is a weather vane, as shown in Figure 10-7, and pretty obviously

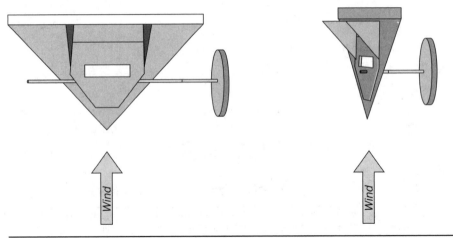

**FIGURE 10-7**    The Flack in a dive showing both pitch and yaw stability.

it would make an excellent weather vane with that shape. If we moved the bearing/CG back to the location in Figure 10-8, we would not get as good a weather vane because there is equal area before and aft of the bearing/CG. If this does not make sense, get out some cardboard and do the experiment. At this point, enough has been explained to make sense of why the CG is forward for both yaw and pitch stability. Our planes do not need to be roll stable because

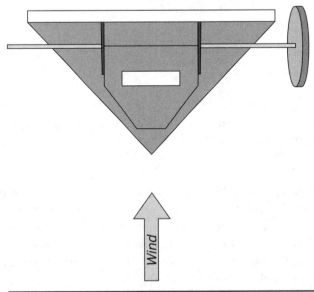

**FIGURE 10-8**    The Flack with an aft CG/bearing demonstrating that the plane will not be stable in pitch.

there are no serious destabilizing roll forces in level flight. At this point, no more learning is necessary to go out there and hack at some sky. But greater mastery can be had by reading on.

### Adding in Lift

We have shown how the location of the CG affects the weather vane stability of the Flack. Essentially, we know how an arrow remains stable in flight or how the plane is stable flying in a vertical dive. Lift makes things very interesting. Lift is an upward force generated by a positive aerodynamic angle of attack that has both the air on top of the wing sucking it up and the air on the bottom pushing it up, as explained earlier. The point at which the lifting forces are balanced is called the *center of pressure,* and for symmetric airfoils, this is located approximately at the dividing line on the wing that has 25 percent of the area in front of it and 75 percent behind, as shown in Figure 10-9.

The CG is one of the factors that helps us make an airplane stable and controllable. A rule of thumb for aircraft design is that the CG of flying wings splits the surface area such that 20 to 25 percent of the wing area is ahead of the CG.

If the CG is ahead of the center of pressure, the plane will have a tendency to raise the nose on dives because the reduced angle of attack will decrease lift, which will lower the tail. On climbs, the effect is reversed because the wing will generate more lift, which will rotate the nose down. This autostabilizing component is how free-flight planes manage to keep flying in turbulent air. With the Flack, the 10½-inch CG placement is slightly autostabilizing in this

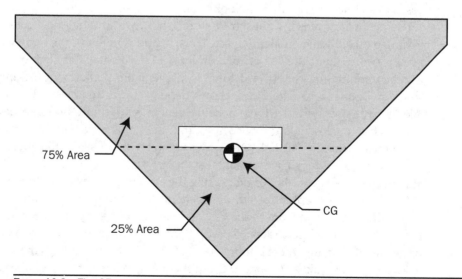

**Figure 10-9**   The 25 percent/75 percent dividing line shown on the Flack that indicates the center of pressure.

way. You can tell by trimming the plane for level flight (see Chapter 5) at two-thirds throttle. The plane should be able to fly 100 feet without any control inputs (it helps if the wind is calm). Place the plane in a shallow dive, and release the controls. You should see the plane gently recover. This is an excellent way to test whether the CG is correct. A slightly forward CG is the preferred setup for beginners because it makes the airplane more docile. It will   also handle turbulence and wind better.

If the CG is at the center of pressure, then the plane will have no tendency to recover from dives or climbing. This is generally how I fly my planes. When trimming a new airplane, I start with the CG too far forward and incrementally keep moving it back until the plane stops recovering from dives on its own. I do this because I value maneuverability over stability.

If the CG is aft of the center of pressure, then the plane can be a real handful to fly. When the nose pitches down, the lift decreases ahead of the CG, and the dive worsens. When the nose pitches up, the lift increases ahead of the CG, and the plane climbs more. Planes with a CG slightly aft of the center of pressure will feel very twitchy in pitch, and they require constant pitch correction. A really aft CG plane is pretty much impossible to fly without onboard gyroscopic intervention.

## Reflex and Flying Wings

You may have noticed that none of the airplanes in this book have a traditional tail like a 737 or a Cessna. They are all flying wings, which are sometimes called *tailless airplanes*. Given that a Flack flies just fine without a tail, why does a Cessna need one? It's not there to look cool.

For a bunch of reasons not worth getting into, wings generate a forward-pitching moment when generating lift. Thus, in addition to lift, the wing wants to roll forward about the CG or tuck under. The horizontal stabilizer exists to provide a countering downward force to prevent the plane from tucking as shown in Figure 10-10. The figure shows the elevator raised a few degrees to make clear that downward force is being generated. Usually no deflection is seen because the downward force comes from the stabilizer's angle of attack to the airstream, which is angled downwards a bit because of the lift being generated by the wing. The horizontal stabilizer also controls the pitch of the entire airplane by increasing or decreasing the amount of downward force.

One way to conceptualize why lift generates a rolling forward force is to assume that the area in front of the CG generates 25 percent of the lift and that the area aft of the CG generates 75 percent of the lift. Since the CG acts like the pivot point of the wing, there is more lift aft of the CG, which explains the rolling forward force. In reality, the amounts of lift being generated fore and aft of the CG vary on many factors, but the general idea is sound.

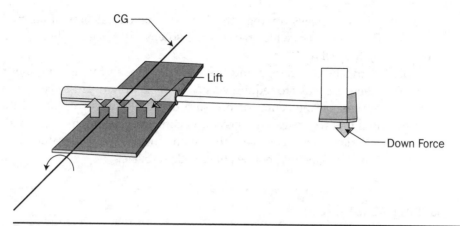

**FIGURE 10-10**   How an elevator uses downward force to counteract the forward-pitching moment inherent in the generation of lift for stable aircraft.

Flying wings, then, are an anomaly because they apparently don't have any horizontal stabilizer to resist this pitching moment—how come they don't pitch forward? Figure 10-11 shows that the elevons are responsible for resisting the pitching moment of the wing by having a little bit of up trim in them. This is called *reflex*. What has happened is that the flying wing does have a horizontal stabilizer; it just happens to be right where the wing ends.

The downward-acting force also helps to keep the airplane stable in pitch. When the nose drops, the angle of attack for the elevator/elevon increases, as does airspeed, which increases the downward force and helps to bring up the

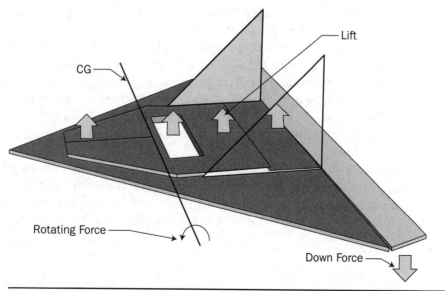

**FIGURE 10-11**   How the elevons on a flying wing counteract the forward-pitching moment.

nose. If the nose pitches up, then the elevator/elevon gets less of an angle of attack and less airspeed, which reduces the downward force and helps to move the nose up to level flight.

Before anyone's panties get into a knot because their flying wing flies with no reflex, there are ways to require very little or no downward force by having the CG be right on the center of pressure or slightly aft of it, but all that is happening is that we are carefully counteracting the pitching moment with the mass of the plane. That balance will become unbalanced when the airspeed changes, and the plane will not be able to self-stabilize.

## Reasoning about Lift

It is worth learning the basic shape of the lift formula if you are designing your own airplanes. It contains some crucial information for the kinds of design decisions you will need to make. A simplified formula for lift is

$$\text{Lift} = \tfrac{1}{2} \times \text{air density} \times \text{velocity} \times \text{velocity} \times$$
$$\text{how "lifty" the wing is} \times \text{how big the wing is}$$

The "liftyness" of the wing is called the lift coefficient and it can be looked up for all sorts of wing shapes and airfoils. For the standard Flack, lift will have to be 15 ounces in level flight, more for climb, and less for descent. What is the crucial information?

1. Velocity is squared, which means that to get twice the lift, you only need to go about 1.4 times as fast.
2. If I want to fly half as fast, the plane will have to either lose 75 percent of its weight, get four times as much surface area, or use an airfoil that generates four times the lift or some combination of the factors. Airspeed makes a huge difference in whether we get off the ground or not.
3. Slightly bigger, slightly heavier is not going to be that big a deal if we can just go a little faster.

How "lifty" the wing is depends on the angle of attack, the efficiency of the airfoil, and the overall shape of the wing. A flat plate is a pretty good airfoil that can generate gobs of lift at the cost of a lot of drag. It is beyond the scope of this book to explore more efficient airfoils, but they are out there if you are interested.

## Drag

Drag is the aeronautical version of friction. It is the force resisting the forward motion of the aircraft. All the airplanes in this book generate a lot of drag because they are simple airfoils and designs optimized for maneuvering rather than efficiency. It is the reason that your Flack flies so badly with party streamers and a balloon tied to it—it's the increased drag. You only need to know that too much drag will keep an airplane from flying well or flying at all, so be aware of it.

## Glide Ratio

This term refers to the distance the airplane goes forward for the amount of altitude lost in a power-off glide. A sailplane can have a glide ratio of 40:1, meaning that for every foot dropped, the sailplane will go 40 feet forward. The Flack's glide ratio has never been measured, but I would estimate it to be between 3:1 and 2:1. This means that the Flack comes down fast, which is a good thing if you are landing in a tight spot with lots of obstructions. It is a bad thing if you smoke the motor over a lake while carrying a camera, and the poor glide ratio causes you to land in the last foot of lake.

## Conclusion

This chapter was meant to introduce the rough shape of what makes a plane fly. It is a big topic. Martin Simons' *Model Aircraft Aerodynamics* is an good start for a more thorough treatment that has gotten respect from serious aerodynamics folks. NASA also has a good web site at www.grc.nasa.gov/www/k-12/airplane/. This chapter hopefully provided enough information to get you into trouble designing your own airplane. Chapter 11 gives you the gritty details of how we do it at the Brooklyn Aerodrome.

# Hack the Flack:
# Make and Fly
# Your Own Design

A benefit of the Brooklyn Aerodrome approach comes from getting to put something strange and fun into the air. This chapter takes you through the process of creating planes out of whole cloth from design to execution. You will have joined the select few who have designed, built, and flown their own remote-control (RC) airplane. It is assumed that standard Flack gear will be used (i.e., motor, servos, etc.) and similar building techniques. There are lots of ways to make airplanes; this is one of them.

The best part is that it is not that hard if you approach it methodically and with a bit of patience. This is not a good place to start as a beginner, but you don't need to be an expert either. Basic building skills and piloting skills are all that are needed. That said, this is the deep end of the book, where you are expected to fill in all the details. I had three testers (thanks Ben, Lowell, and Andrew), and it went okay, but not rock solid like the first five chapters. This chapter lays out basic frameworks for novel flying wings. It is not going to work well for traditionally laid-out designs or canards—so it is limited (I find that I do my best work with constraints).

Two major approaches to novel aircraft are presented. One is incremental, based on starting with a small glider and working your way to a powered RC aircraft. The other approach works by removing material from known good designs to achieve the design objectives. And remember to send me a picture/video at bible@brooklynaerodrome.com.

## Getting the Idea

What gets your creative juices flowing? Do you want a night flyer? Do you want something that looks like an airplane or not at all like an airplane? I constantly seek inspiration from the animal and vegetable kingdoms. The flying-wing approach I pursue is well suited for big creatures, abstract shapes—How about some flying lips?

## Prototyping

Once you have some ideas, the key issues are whether the shape can generate lift and be controllable. There are computer programs out there that can reduce the risk of a new design by modeling the lift, control, and structural issues around your idea, but I have never used them because there is a simpler way—build a half-scale model glider and see if you can get it to fly.

Figure 11-1 shows some of the gliders I have built when trying out new ideas. Blue foam is an excellent material to "sketch" with, and I use subway cards as stabilizers. Getting a glider to fly can be a little tricky because you may not know what the center of gravity (CG) should be or appropriate trim. Below are some steps you can take to maximize the chance that your glider can be made to fly. The first glider to make is one with known properties that are easy to work with—the classic plank flying wing.

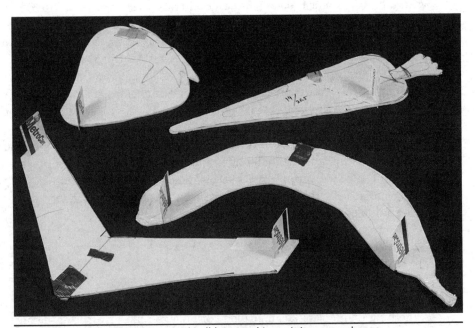

**Figure 11-1**  Various 50 percent scale gliders used to prototype new shapes.

## A Plank

I strongly recommend that you build a plank to help calibrate your expectations as to how to launch, trim, and evaluate prototype gliders. It will take 5 minutes of your time once you have the materials. The steps include

1. Cut a 12- × 8-inch plank from foam or whatever you are using.
2. Bend up the last 1 inch of one edge to have ¹⁄₁₆ inch of up trim for reflex. This will be the trailing edge. A good way to do this is to place the wing on the edge of a table with 1 inch hanging off the edge. Take a ruler, and use it to force the bend on the 1-inch section.
3. Split the elevons in the midpoint, and keep cutting farther so that a business, playing, or subway card can be inserted vertically into the slit.
4. Add the business, playing, or subway card as a stabilizer in the center slit.
5. Launch the plane. Observe how it flutters to the ground in a very nonflying way. This is a plane with a very far aft CG. The next steps will predict where the CG should be.
6. Measure 2 inches back from the leading edge, and mark both the top and bottom with a line parallel with the leading edge. This line is the 25 percent chord point of the wing. Chord is the dimension of the wing that is geometrically parallel with the direction of flight. Chord does include elevons.
7. Flying wings work well if the CG is between 20 and 25 percent of the mean aerodynamic chord (MAC) of the wing. Twenty-five percent MAC is the point at which fore and aft split the surface area of the wing 25 percent/75 percent relative to the direction of flight. For the plank design of this example, the calculation is trivial—I just did it by measuring from the leading edge. For strange shapes, MAC can be quite difficult to determine.
8. Add sufficient weight at the nose to have the plane balance at the 25 percent chord point. Put the weight on the bottom side to give your finger a place from which to launch. The plane in Figure 11-2 balanced with three quarters and two pennies.
9. Launch the plane with a firm flick of the wrist, level in both pitch and roll. Or you can launch like you learned with the Flack. Do not throw it up. Level launch is the goal.
   a. If the glider pops up, then reduce up trim a little.
   b. If the glider dives, then increase up trim.
   c. If the glider turns left or right, adjust with opposite elevon trim.
10. Keep at trimming the glider and improving your launch technique until the glider is flying at least 20 feet. This glider is your reference for trimming and evaluating future gliders. Some gliders will fly worse,

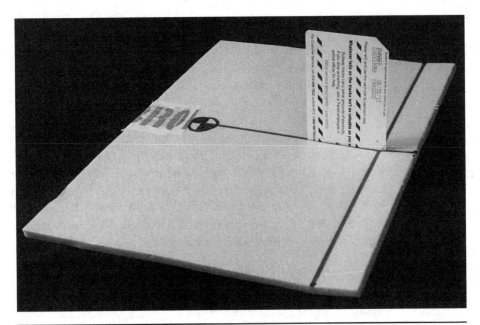

**Figure 11-2** Plank flying wing with ¹⁄₁₆-inch up trim and nose weight.

some better, but the plank is helpful for assessing likely performance and viability of your idea.

## Designing and Building a Novel Design

Building your own design glider is a little more open-ended because I can't possibly know the degree of your genius. But this is how I proceed:

1. Draw the initial design in top view on graph paper with a 24 × 48 grid outline. Scale it to what you think will be a 50 percent scale glider, and cut it out of foam. The grid will really help with scaling to a full-size model.
2. All flying wings require some reflex or up trim to be self-stabilizing in pitch. Figure 11-1 shows how I have bent the trailing edge up a little bit on various shapes. Sometimes an elevon needs to be explicitly cut, or the trailing edge can be just bent up as with the plank.
3. Add a stabilizer as far back as possible from the leading edge or where the design calls for it.
4. If possible, determine the point on the wing that splits the wing area 25 percent/75 percent in the direction of flight. Add enough weight to the

nose to achieve that weight distribution. I just pick a point one-quarter of the way back on the chord of the wing and add weight until the glider balances on my fingers at that point.

    a. Launch the glider with attention to elevon adjustments for flight path.

    b. If elevon trim is not working or looking too extreme (see reference plank build), then adjust nose weight in very small increments (one penny) as follows:

       i. If the glider is diving, then remove nose weight.

      ii. If the glider is climbing and stalling, then add nose weight.

    c. Keep working with nose weight and elevon trim to optimize stability and glide path.

    d. If you like what you see, consider building a full-size version.

5. If the shape is too complex to determine the CG point (e.g., the Banana or Carrot), then apply the "that looks about right" (TLAR) method of adding nose weight. I have never built a glider that did not need nose weight. This makes it harder to know whether the CG is correct, so I fix the elevon trim at something believable (again, consult the reference plank) and add or remove nose weight to figure out what the CG should be in one-dime increments.

    a. If the nose is rising, then add weight to the nose.

    b. If the glider is diving, then remove nose weight.

    c. If the CG adjustments are not working, then adjust elevons further. Keep trying things until you get the glider flying well.

    d. If you like what you see, build it full size.

6. Assess the directional stability of the glider. If it is falling off left or right, add more stabilizer area to try to keep it flying straight. Remember that this may move the actual CG away from the design CG.

7. Be forewarned that actual glider CGs tend to come out too far aft for an RC airplane. Factor this into your anticipated weight distribution in the full-size build.

## Building the Powered Version

If you have a glider that flies about as well as the plank, then there is an excellent chance that a powered version will fly as well. The next step is to size your creation. But please read this entire chapter before proceeding. Many factors influence scaling to a powered prototype that have to be considered simultaneously, and books are inherently linear.

## Scaling Up from the Glider Proof of Concept

This book assumes that the same basic materials that I used for the Flack will be used. These include foam, speed control, motor, servos, and battery.

### Wing Area

The Flack is sufficiently powered and has enough wing area that almost any shape you can get flying as a glider likely will work if you stick with the wing area and overall weight of the Flack. An easy way to calculate wing area is to draw the design in SketchUp, select its surface, and use the Entity Info menu, which will pop up a window showing the surface area of the design.

Another way to measure wing area is to draw a 2-inch grid on the design and count how many squares there are. For partial squares, just guess how much of a square there is—the measurement does not need to be exact.

The Flack has 425 square inches of wing area, including elevons, so try to size your design to match that. Significantly more wing area (100 square inches or more) will make the airplane slower, assuming no changes in weight. Significantly less wing area will make the airplane faster. Look at Chapter 8 for an idea of how surface area affects performance across the designs.

Also consider the natural shape of your raw materials. Blue foam comes in 24- × 48-inch rectangles. Keeping wing spans under 48 inches will mean that joints won't be needed.

### Weight

The Flack is very happy with a flying weight between 15 and 22 ounces. But the delta-wing design generates lots of lift, so don't assume that you have the same weight-carrying capacity because you have the same wing area. If your design is very "un-wing-like," then a good move is to keep everything as light as possible. It is hard to define what "un-wing-like" means, but some examples include

1. A flying superhero—this has actually been done.
2. Flying the carrot pointy-tip first.
3. A flying doily with lots of holes in it.

If the test glider flew less far than the plank glider by 50 percent, then you likely have a shape that is "un-wing-like."

That said, almost any flat shape will generate some lift, so keeping it light increases the chances of success. The minimum weight possible is around 13 ounces for Flack-class designs just because of the equipment and foam. Often what makes an airplane heavy is the weight added to achieve a balanced CG. Also, careful placement of the battery and motor can really help to keep the overall weight down.

### Control

The glider prototype should have forced the design into being statically stable, but this doesn't mean that there is a way to control the aircraft. Many a design has failed because I could not work a way to put elevons on it. Some failed efforts include

1. A flying tadpole
2. A flying candy cane with handle forward

This is not to say that it cannot be done, but that it is challenging.

### Sizing Elevons

Chapter 9 is a good place to get a sense of how elevons should be designed on novel shapes. Keep the area similar to a Flack's elevons, which are approximately 100 square inches total. Another metric is to have the elevons be one-fifth the surface area of the entire wing. Reasons to have bigger elevons include the elevons not being in the prop blast, which reduces control authority, and having elevons that are oddly shaped.

Elevons also can be too big, which can result in servos that are overwhelmed with the aerodynamic forces on them or an airplane that is overly twitchy on the controls. If you don't occasionally use full control throws (e.g., landing, launching, and acrobatics), then consider making the elevons smaller if the control throws are at 45 degrees or less.

### Sizing Stabilizers

I do not test gliders to evaluate stabilizers generally. I use an expired subway card to provide yaw stability and do my serious thinking when the full-size design is being built. The Flack flies just fine with one stabilizer, and with deft piloting, it can be flown with no stabilizers at all. Generally, stabilizer size is more than sufficient for aesthetic reasons.

The Flack with a single stabilizer uses 32 square inches starting 2 inches from the CG and ending 10 inches away to provide yaw stability. The Flying Heart uses a single stabilizer starting 12 inches from the CG with a surface area of 60 square inches—it has more stabilizer area than probably is needed, but it looks good.

The one case where stabilizer area gets tricky is if there is a significant destabilizing vertical area ahead of the CG. Figure 11-3 shows a Flying Tadpole prototype that had a large dome in front of the CG. That surface area had to be compensated with a sizable vertical stabilizer for stability, as shown. The size of the stabilizer was determined by taking the vertical area of the Tadpole body and making the stabilizer 1.5 times that area. During flight testing, the stabilizer area was reduced gradually until the plane became slightly unstable. The area of the last stable configuration was used.

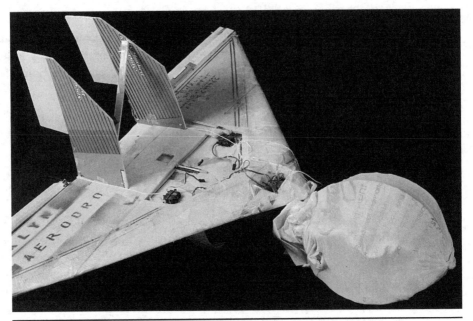

**Figure 11-3**   Flying Tadpole with big destabilizing nose and compensating tail.

## Structural Considerations

Once the basic airframe has been cut out and the elevons cut and hinged, it is time to decide whether reinforcement is needed. Some type of reinforcement almost certainly will be needed for the motor mount and servos. If the final airframe will be stiffened by covering, the prototype can use Coroplast as a proxy stiffener. There are no hard and fast rules, and experience dominates—I have made many overly floppy prototype airplanes.

Build with strength in mind. My prototypes crash pretty hard about half the time on first launch. Generally, this occurs because I got the CG entirely wrong. I am getting better about this, but it still happens. Coroplast is your friend in such situations.

### Motor-Mount Attachment Points

The motor generates a lot of force on the motor mount when flying and particularly on landing. It needs to be able to take abuse. The standard motor-mount attachment from the angle stock to the airframe is not always possible, however. I have sandwiched the motor mount between two sheets of blue foam and zip ties, as done with the battery, or used high-stick tape to attach it to adhesive plastic film.

### Servo Attachment Points

Servos need solid mounting as well. High-stick tape on plastic film can be strong enough on its own to hold servos down.

### General Stiffening

The standard stiffening agent for blue foam is either a layer of Coroplast or a layer of plastic film. Both work really well to stiffen, and they provide a solid surface for motor mounts and servos. Another route for stiffening was used for the Bat, as shown in Figure 11-4. It was doubled blue foam, and it worked very well and has endured many flights, firm landings, and transportation. Packing tape was used to stiffen the opposite side of the motor mount. Use the Flack's stiffness as a guide to determine whether your design is stiff enough. Long, thin wings are going to need some help. Look at the shapes in Chapter 9 for guidance on what does and does not need stiffening.

### Placing Equipment

It is a very good idea to experiment with different placements of equipment to attempt to achieve the expected CG as determined by the glider. The heaviest items are the most useful for this, and these include the battery (3.4 ounces),

**Figure 11-4**  Details of both foam doubling for stiffness and short-rod servo installation on the Bat.

followed by the deck (2.5 ounces), with the motor/prop/motor mount coming in at 2.1 ounces. The rest of the components are less than an 1 ounce each, so they are less likely to have a big impact on airplane balance. Remember that the glider has determined the design CG that needs to be achieved.

Almost all of our designs at Brooklyn Aerodrome would fly better with the motor on the front, but our desire for safety means that the hard metal bits need to be surrounded by foam. Please keep safety a major component of your designs as well.

## Flight Testing

Nothing could be finer than giving a new design a huck into the wild blue yonder. These are the steps for a new design's first flight:

1. Do yourself a favor and verify that surfaces move the right way both in the studio and on the field before your first launch. The 3D Banana was pounded into the ground by reversed aileron control.
2. Verify that the CG mirrors that of your glider. Make it 1 inch forward of that location because the gliders tend to have aft CGs. A plane with a too-far-forward CG is flyable, whereas a plane with a too-aft CG tends to be uncontrollable.
3. Have ¼ inch of up reflex with your elevons.
4. Have someone who knows how to launch launch the plane. Have that person practice on a Flack if they are not sure.
5. Consider different ways of launching the plane. The Banana planes are both launched from the middle with an underhand toss.
6. Fly over tall grass if at all possible. It really cushions crashes.
7. There are two schools of thought about whether to apply power on first launch. It depends on the wing loading of your design, how fragile it is, and how soft a spot it will land on. If there is knee-deep wheat fragrantly wafting before me, I will do my first launch without power if the design has a remote chance of gliding. Here in Brooklyn with hard-scrabble dirt or asphalt, I always launch with power—this generally results in a crash, which is why prototypes need to be built tough. The steps for each approach are as follows:
   a. An unpowered launch needs enough airspeed to test the aerodynamics. Launch just like you did when learning to fly the Flack with a pilot on the controls. Level, firm, and set to land 10 feet out. Make adjustments and repeat until that throw is controlled. Then move to the powered launch approach.

  b. For a powered launch, the first rule is to launch at full throttle. I have failed to do this and crashed. Keep at the launching, making adjustments based on the following.

8. Fly the plane like a beginner. Just get level flight to 50 feet out and land. Then start pushing maneuvering.

9. If the plane has anything wrong or is difficult to control, then cut power as much as possible, and get the plane down.

10. Be ready with up elevator. If it is very hard to get the nose up, then consider
  a. Moving the CG back ½ inch
  b. Making the elevons bigger
  c. Getting a more powerful motor

11. Be ready for the airplane to be very squirrelly in pitch. If it is, then the CG is likely too far back. Move it 1 inch forward.

12. Be ready for the airplane to tuck or dive uncontrollably. This can be from a too-far-back CG or an airframe that is too soft and curving in flight. The latter produces a huge pitching-forward moment that can overwhelm the elevons, as happened with the 2D Flying Heart.

13. Video the flight tests. It can be very useful to have a record of what actually happened. In addition, if it works, you can put it on YouTube.

14. Don't fly too much the first time out. Go back to the lab and think about how the airplane is flying.

15. As you get to know your new design, try small changes once you have it roughly controllable and flying. Move the CG ¼ inch in various directions to see what it does. Increase/decrease elevon throws, stabilizer sizes, and elevon sizes.

## New Shapes by Morphing Old Ones

Figures 11-5 and 11-6 are examples of creating a new shape by slowly morphing a known shape into the desired shape incrementally. It is an excellent way to adapt to extremely radical designs. The Flying Chandelier started out as a traditionally laid-out plank wing with a tractor motor. It flew well, as expected.

### Incremental Refinement

Next, the top and bottom edges of the Chandelier arms were cut out on the wings, and the airplane was test flown with the CG moved back a bit to compensate for the loss of wing area. Flight performance was slightly degraded but acceptable.

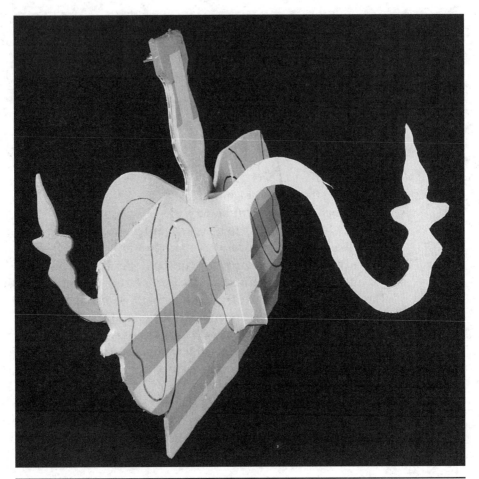

**Figure 11-5**    Flying Chandelier with planned wing outlines in electroluminescent wire.

The Chandelier needs four arms, so the vertical stabilizer was fashioned out of foam, and a bottom arm was made out of Coroplast to withstand being landed on. This version was flown successfully as well with increasingly degraded aerodynamics.

The goal was to make this airplane a night flyer with electroluminescent wire (el-wire) outlining the shape of the arms. The wings eventually would have had clear material where indicated in Figure 11-5. Flight testing pointed out that the design needed to be made bigger to slow it down and make it more controllable. Ultimately, this design was abandoned for lack of interest, but the strategy of incremental refinement worked perfectly and pointed the way to needed design changes as well as validating the overall idea.

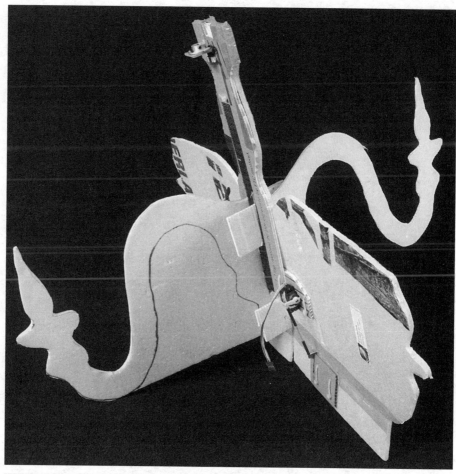

**FIGURE 11-6**    Servo and other gear installation shown for the Flying Chandelier.

## Conclusion

This chapter ties together the sorts of knowledge and techniques needed for creating your own designs. It is very fun and rewarding—there is nothing like launching a completely new shape with weird aerodynamics and seeing it work. After my first test flights of the Flying Heart, Bat, Manta Ray, and many others, my hands shook from the excitement of the event. I have to keep first flights short because I am afraid that my nervous thumbs will crash the plane. It is an awesome experience to have. Send me a picture/video of what you come up with at bible@brooklynaerodrome.com.

CHAPTER **12**

# Simulators, Autopilots, Video, and Buddy Boxing

T his chapter discusses additional equipment that can be fun or helpful in your plane-building experiences. Some areas, such as autopilots and live video feeds from the aircraft, are beyond the scope of this chapter, but the current state of the art is discussed as well as where to get more information.

## Flight Simulators

An excellent way to learn to fly is with a flight simulator. There is a broad range of remote-control (RC) flight simulators available at price points from free to hundreds of dollars. The free ones are perfectly adequate for getting the basics of flying sorted out. I strongly recommend getting a USB RC transmitter emulator to drive the software. The USB controllers interface effortlessly with Mac OS or Windows as a standard USB game controller (see Chapter 1). Using a game controller or mouse will interfere with the learning. USB flight-simulator controllers can be had for around $20 from hobbypartz.com as of this writing, but check out this book's website for updates. In addition, there may be a way to use the transmitter you have to drive the simulator—check your transmitter's manual.

---

WARNING: *The Dynam USB simulator controller has a Mix switch that makes a mess of the control inputs if it is on. Make sure that the Mix switch is off.*

---

### Flying Model Simulator (FMS)

FMS is a free flight simulator that only runs under Windows. Figure 12-1 shows it in action. Thanks to Gary Gunnerson, a model of the Flack is available for the simulator (http://gunnerson.homestead.com/files/towel.zip). Visit http://

**Figure 12-1**   A Flack in FMS.

modelsimulator.com/ and download the relevant version of the software. For almost all of you, that will be Version 2 alpha 2.5 unless you are sporting a Windows 95 box. Any Windows version more recent than XP needs a library added. Look at the instructions at http://www.microflight.com/s.nl/ctype.KB/it.I/id.1390/KB.1034/.f, and you can download the .dll at http://www.microflight.com/img/d3drm.zip.

## CRRCsim

Charles River Radio Controllers Simulation (CRRCsim) was developed by Jan Kansky and Mark Drela out of Massachusetts Institute of Technology. Professor Drela teaches aeronautics and has been a major innovator in the world of RC gliders and human-powered flight. Given the credentials of the authors, I assume that the physics modeling is excellent.

CRRCsim runs on Windows, Mac OS, and LINUX. Look for precompiled distributions for Windows and Mac OS. A tricky part of the interface is that to access any of the settings, the Escape key (ESC) has to be pressed.

1. Connect your USB controller, and launch CRRCsim. Press the ESC key, and click on Options and click on Controls.

2. Select the Input Method, and choose Joy Stick. Go through the Configuration process, and assign up/down on the right stick to elevator, left/right on the right stick to aileron, up/down on the left stick to throttle, and left/right on the left stick to rudder. You may need to invert some of the controls so that they move the simulation the right way. Figure 12-2 shows the control settings for the Dynam USB controller. Pulling back on elevator should go up, aileron should roll the model the correct way, and throttle is zero when the left stick is down and 100 percent when it is up. You will have to select a powered airplane from the Options > Airplane > Super Zagi before verifying the controls (Figure 12-3).

The airplane is launched with the "R" key or Simulation > Restart. There are many settings available, and it is good to explore them.

When the simulator is set up, the most similar airplane to the Flack is the Super Zagi. Practice flying it in figure eights, try to fly it at your head, and then fly it inverted. Do the exercises in Chapter 4 for learning to fly with the added benefit that you don't have to fix anything after a crash. There may be compatibility issues with CRRCsim in the latest versions of OSX. Visit this book's website for updates.

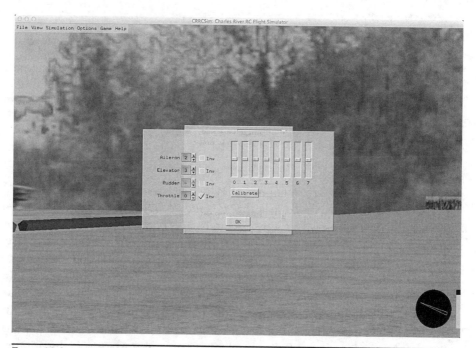

FIGURE 12-2   Configuration for the Dynam USB controller.

**Figure 12-3**    Electric Zagi flying in CRRCsim.

## Autopilots

There are lots of ways to make your airplane more autonomous, ranging from autostabilization systems to full-on autopilots that will do GPS waypoints. This section starts with a commercially available autostabilization system and then provides an overview of full-on autopilots.

### Copilot CPD4

The FMA/Revolectrix Copilot CPD4 rather brilliantly rights a plane to level pitch and roll when the control stick is centered. When the plane is out of control, the pilot just releases the right stick, and all is good. I have had complete novices fly for minutes at a time without any intervention from me. It is also a very nice security blanket for night flying. This device costs around $60 and works very well. The technology works by a four-window sensor that measures infrared (IR) energy, and it turns out that the ground tends to be much warmer than the air. In level flight, the sensors should report the same IR in all four directions. If one direction is "hotter" and the opposite sensor is "colder," then that indicates that the airplane is not flying level, and the computer applies the appropriate inputs to correct the situation if the controls are centered. Quite clever. The CPD4 manual is quite thorough, but because it handles helicopters, traditionally laid-out aircraft, and flying wings, it can be a bit of a challenge to work out exactly what to do. Do download the manual at www.revolectrix.com/support_docs/item_1126.pdf.

If you are a beginner without any help, use of the CPD4 is not recommended by FMA. Chapter 5 works really well for learning. If you try to learn on your own using the CPD4, please share your successes/failures with me at bible@brooklynaerodrome.com. I will keep this book's website up-to-date with what I learn.

## Full Autopilot

The equipment required to have an autonomous drone has shrunk in size and cost so much that small RC aircraft can be the foundation. The kinds of aircraft in this book are not that well suited to this role because they have limited duration and range, and there is no safe place to put all the electronics. But it really would be fun.

---

Model aircraft operation without line-of-sight access is most likely illegal or will be soon when the Federal Aviation Administration releases new unmanned aerial vehicle (UAV) regulations. It is also generally understood that model airplanes fly no higher than 400 feet.

---

The current state of the art in the hobby autopilot world is as follows:

1. Aircraft are launched and landed under remote control. This may be very easy to do given that there will be an autostabilization system such as the CPD4 on board.
2. There is a full autopilot that can maneuver the airplane.
3. There is GPS tracking, and it is integrated with the autopilot.
4. There is flight-planning software that programs the autopilot with waypoints.
5. Real-time data can be streamed from the plane to the ground. Those data can include live video, operational parameters, and locations.
6. Radio ranges can be beyond line-of-sight. Note that beyond line-of-sight operation is likely illegal or will be soon.

There is a vibrant community constantly improving on the state of the art. An excellent source of information is the website DIYDrones.com.

## Video from the Air

I have been putting video cameras in the air for years and having a great time doing it. There are two major approaches to video: (1) passive video, which puts a camera up there and records what comes by, and (2) first-person video (FPV), which streams to a base station that the pilot can use to fly the airplane, record, etc.

### Passive Video

Figure 12-4 shows my favorite camera that I have strapped to airplanes and put in harm's way. Go to the Brooklyn Aerodrome YouTube.com channel to see some of what we have shot. We typically mount the camera on the bottom of the wing as close to the CG as possible to keep the balance of the aircraft. This is our preferred mounting method, and it makes clear why the cameras take a beating on landing.

Some lessons we have learned shooting passive video:

1. Wide-angle lenses work better than narrow-angle lenses.
2. Flying really pushes the limits of onboard video compression. Make sure that the camera can handle rapid scene changes without introducing artifacts into the video signal.

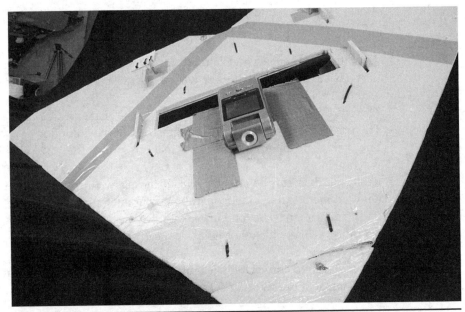

**FIGURE 12-4**   Sony Webbie ready for another round of abuse. So far it has been dropped from 250 feet, immersed in Prospect Park Lake, and run over by a truck.

3. Learn to fly long, straight paths over the subject. Lots of turns are disorienting.
4. We try to point the camera down about 30 degrees toward the front.
5. Windy days make for rough video footage.
6. Velcro can and has released. Make sure that the camera cannot fall off the plane with a safety leash or zip ties.
7. Bring spare storage cards or a computer to download photos if the video is important. Planes get caught in trees, land in water, and get run over by trucks, and you don't want to lose previous flights video.

The great thing about passive video is that it is cheap and easy to do. Strap your cell phone to a plane and have at it. Great fun. The not so great thing about passive video is that the videos are generally not that good because of framing issues, bad flight paths, and either being too high or too low. The solution to these problems is with FPV in the next section.

## First-Person Video (FPV)

I started to explore FPV with Andreas back in 2006 when it was just starting to gather momentum. It was a huge hassle to select gear and get it to work well, and I lost interest because of the complexity. Things are all grown up now, and complete FPV kits can be had for $300.

At its simplest, an FPV setup requires a video camera on the airplane, a transmitter on the airplane, a receiver on the ground, and a video monitor on the ground. The ground monitor can be video glasses. Figure 12-5 shows the apparently no longer available Fly Cam One basic FPV setup.

Depending on setup, there may be a higher-quality camera on the airplane oriented the same as the pilot's camera. These rigs can get expensive fast, and one has to wonder about the wisdom of putting all that expensive gear on an airplane that is made to crash. But I have done it.

FPV brings some interesting capabilities to the table:

1. It allows much better video framing than passive video.
2. It is not necessarily easier to fly than regular RC. It is very easy to lose orientation with the narrow camera view.
3. Non-line-of-sight flying is possible but likely illegal.
4. It is hugely fun.

As with autopilots, there is a rich world of information out on the Web about FPV. Anything I write here will be out of date within 6 months anyway. A good site to start with is www.fpvpilot.com.

**Figure 12-5**    Fly Cam One looking good while it looks at you. Base station and airborne unit are shown.

## Buddy Boxing a.k.a. Student Driving for Pilots

When I go to Maker Faires with lots of open space, I like to offer buddy-box sessions, where four students listen to a quick training session about how the airplane works and a brief introduction to the transmitter controls. Then I launch a plane and let the pilots try to fly. Three mistakes, and the next student pilot takes the transmitter. I cycle through until the battery is depleted. It's a blast.

A big part of the success of this program is that the buddy-box setup I use allows me to give elevator control without aileron, aileron without elevator, or both controls to the student. It is much easier to control a single dimension of the airplane when learning. A surprising number of students learn in a few minutes to control both pitch and roll with this process.

Lots of RC transmitters allow for a teacher transmitter to connect to a student transmitter, with the teacher controlling who has control of the airplane. The Tactic discussed in Chapter 1 has a particularly nice version that is a wireless setup. The Tactic, however, gives complete control to the student, which can be overwhelming. This section covers how to hack a pair of radios to give selective control as described earlier.

## Hacking the Transmitters for Buddy Boxing

The student control only allows control of the elevator and aileron. The throttle is taped off to make clear to students that they have no control of the throttle. The teacher control has been modified to have the two switches on top of the transmitter control whether the student has elevator control and/or aileron control. The forward switch is student control; backwards, teacher control. In my experience, it is faster to pull back on switches to recover control, and sometimes the teacher needs to be really fast. The example I am working from is a pair of HKT6a transmitters.

The hack is driven by the switches determining which potentiometer is serving as input to the transmitter radio frequency (RF) section. The circuit diagram in Figure 12-6 shows the basics. Stupid simple, eh? A minimum of four wires need to connect the student to teacher transmitters. Ethernet cable is a readily available source of high-conductor-count wire, and if you want to get fancy, you can put Ethernet jacks on each controller. This hack most definitely voids the warranty on your radios and destroys the student radio entirely. Circuits and components change all the time, and for that reason, I am not providing detailed step-by-step instructions with photography. If you don't understand the circuit diagram, then find someone who does. But I will show you the mess I made, and it works great.

The guts of the transmitters are up to you to sort out—they all differ. The circuit diagram that embodies the idea is given in Figure 12-6.

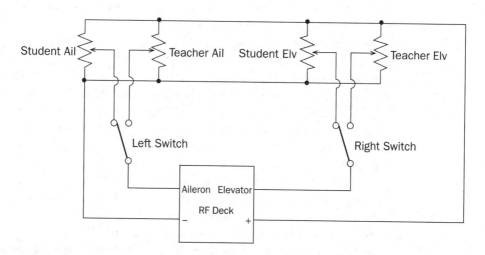

**Figure 12-6**  Circuit diagram for allowing student or teacher control of aileron and/or elevator. This is a setup that maximizes learning to fly quickly.

You will need a Phillips screwdriver, a soldering iron, heat-shrink tubing, and at least 6 feet of Ethernet cable or other at least four-conductor cable. The high-level steps to create the buddy box setup include

1. Open up the transmitters and put them on something that will not push on the sticks too hard when you are working on them. The box they came in likely will work great. Remember that left and right switch so that the elevator/aileron controls are on the left instead of the right.

2. For the elevator, I use the right switch (remember, it will be the left switch because we are looking from the back) to determine whether the student or teacher controls the airplane. It turns out that the broad class of radios based on the Fly Sky FS CT-6B uses a fancier switch than strictly necessary. It is a single-pole, double-throw (SPDT) switch that allows for selection of which of two wires completes a circuit. One circuit is the student's elevator input; the other the teacher's. If you cut the wires at the motherboard for the switch, you may be able to save yourself some soldering.

3. Cut the teacher's elevator potentiometer output from input to the RF deck. It is the middle wire coming off the potentiometer board. Take the middle wire from the switch and solder it to the elevator input wire on the RF deck—find the wire by following the middle wire to where it attaches to the motherboard. It will likely be a wire-wire solder joint. Then connect the outside wire on the switch to the elevator potentiometer solder pad. Now, when the switch is back toward the teacher, the teacher is in control of the elevator. Verify that this works. With the switch forward, there is no elevator control.

4. Do steps 2 and 3 for the aileron, substituting the left (right when working on the back of the transmitter) switch for the aileron output. Now, with both switches back, the teacher should be in full control. With the switches forward, there is no control yet.

5. Drill holes in the appropriate parts of the transmitter body of the student and teacher transmitters, and route Ethernet or whatever four-conductor wire you have through after cutting off the heads. If you keep the holes just a bit bigger than the wire bundle, you can put stress-relief zip ties to keep the solder joints you are about to make from getting ripped apart when the cable gets a yank.

6. Assuming use of Ethernet cable, double up all the wires, reducing the eight conductors to four. Four connections need to be made; use your skills to plan wire length to keep it neat.
   a. Solder one conductor to the remaining tab on the elevator control switch. You may need to remove the switch from the transmitter housing to get access to the other terminal.

b. Solder one conductor to the remaining tab on the aileron control switch.

c. Solder another conductor to any positive pad on the control sticks. Use a multimeter to confirm this. Solder the other conductor to the corresponding negative pad.

d. On the student transmitter side, cut all outgoing connectors out of the elevator and aileron pots. Leave sufficient length for the positive and negative leads of one to be soldered to the pads of the other pot. Solder the power conductors to the pads of one of the pots—this gets plus (+) and minus (–) to the student pots.

e. Solder the aileron conductor to the aileron pot pad and the elevator to the elevator pot pad.

f. You now should have student inputs to the RF deck with the switches forward, teacher with them back.

## Conclusion

Almost everything in this chapter is constantly evolving. For example, a web search on "RC flight stabilization" produced two systems I had not known about. The legal landscape for autonomous flight is also certainly going to change soon, so stay informed.

That's it for this book. This is the book I wish I had when I started out in 2005 with the initial Brooklyn Aerodrome project—well, maybe not since a big part of the fun was learning and making mistakes along the way. I hope you take this as a starting point and push the state-of-the-Flack to higher levels. Send e-mails, videos, and comments to bible@brooklynaerodrome.com. It makes my day every time I learn of someone building a Flack, learning to fly, or being creative.

# Index